**Venice in Las Vegas**

# Studies in Criticality

*Series Editor*
Shirley R. Steinberg

Vol. 562

Alan N. Shapiro

# Venice in Las Vegas

An American and European
Auto-Socio-Biography, 1960s to 1980s

# PETER LANG

New York · Berlin · Bruxelles · Chennai · Lausanne · Oxford

Library of Congress Cataloging-in-Publication Data
Names: Shapiro, Alan N. (Alan Neil), 1956- author
Title: Venice in Las Vegas : an American and European auto-socio-biography,
   1960s to 1980s / written by Alan Shapiro.
Description: Lausanne ; New York : Peter Lang, [2025] | Series:
   Counterpoints, 1058-1634 ; vol. 562 | Includes bibliographical
   references.
Identifiers: LCCN 2025015847 | ISBN 9783034353649 paperback | ISBN
   9783034358552 ebook | ISBN 9783034358569 epub
Subjects: LCSH: Computer scientists--Biography | United States--Social conditions
   | Europe--Social conditions | LCGFT: Autobiographies
Classification: LCC QA76.2.S544 A3 2025 | DDC 004.092 $a
   B--dc23/eng/20250611
LC record available at https://lccn.loc.gov/2025015847

**Bibliographic information published by the Deutsche Nationalbibliothek.**
The German National Library lists this publication in the German
National Bibliography; detailed bibliographic data is available
on the Internet at http:// dnb.d-nb.de.

New York: Michael M / Unsplash
Venezia: Alex Vegas / Unsplash
Roulette Table: Adobe Stock
Cover design by Peter Lang

ISSN 1058-1634 (Print)
ISBN 978-3-0343-6247-4 (paperback)
ISBN 978-3-0343-5855-2 (E-PDF)
ISBN 978-3-0343-5856-9 (E-PUB)
DOI 10.3726/b22883

© 2025 Peter Lang Group AG, Lausanne, Switzerland
Published by Peter Lang Publishing Inc., New York (USA)
info@peterlang.com – www.peterlang.com

All rights reserved.
All parts of this publication are protected by copyright.
Any utilization outside the strict limits of the copyright law, without the permission
of the publisher, is forbidden and liable to prosecution. This applies in particular to
reproductions, translations, microfilming, and storage and processing in electronic
retrieval systems.

*To the memory of Paul Auster.
I cannot write as well as you, but you inspired me to write my story.*

*A group of books on a table (Photo by Alan N. Shapiro)*[1]

# Table of Contents

| | |
|---|---|
| List of Illustrations | 9 |
| Foreword | 13 |
| The Beginning | 17 |
| Acknowledgments | 19 |
| 1    When I Was Twenty | 21 |
| 2    The Scene of the Crime | 37 |
| 3    The Radical Left Late 1960s | 51 |
| 4    High School | 77 |
| 5    Massachusetts Institute of Technology | 91 |
| 6    Cornell University Arts and Sciences | 105 |
| 7    First Year in France and Italy | 125 |
| 8    Art Students Make Politics: the "Metropolitan Indians" in Italy | 135 |
| 9    Back in America Only to Leave Again | 153 |
| 10  Two Years in Bologna | 173 |

TABLE OF CONTENTS

| | |
|---|---|
| 11  Las Vegas in Venice | 189 |
| 12  Sociology Graduate Student in New York City | 207 |
| 13  Greenwood Mills Marketing Company | 231 |
| 14  Wall Street Computer Programmer | 241 |
| 15  Venice in Las Vegas | 255 |
| Epilogue | 273 |
| Notes | 277 |

# List of Illustrations

| | | |
|---|---|---|
| 1 | Books by Paul Auster on a table | 5 |
| 2 | Gondola at Venetian Resort Hotel Casino, Las Vegas | 14 |
| 3 | Ca' d'Oro Palace in Venice and in Las Vegas | 15 |
| 4 | Ceiling Painting at Venetian Resort Hotel Casino, Las Vegas | 16 |
| 5 | Alan as a baby | 25 |
| 6 | A handwritten list of structural engineering projects made by my father shortly before his death | 29 |
| 7 | My parents, Murray and Florence Shapiro, in the early 1950s | 30 |
| 8 | The model house of our cookie-cutter house, Williston Park, LI, NY, 1958 | 31 |
| 9 | Painting "Connecting Fences" by my mother Florence Morrison | 36 |
| 10 | Sands Point Country Day School mansion at Elm Court | 39 |
| 11 | Members of the Kennedy family leaving the U.S. Capitol after viewing JFK lying in state | 46 |
| 12 | Painting "Fishing Village" by my mother Florence Morrison | 49 |
| 13 | My father and I: 1966 Sands Point Country Day School sixth grade graduation | 50 |
| 14 | Steve McQueen on a motorcycle in *The Great Escape* | 52 |
| 15 | Treasure Island Hotel-Casino Las Vegas | 55 |
| 16 | Herricks Junior High School Math Award, 1969 | 56 |

## LIST OF ILLUSTRATIONS

| | | |
|---|---|---|
| 17 | *Star Trek: The Original Series*, episode "Shore Leave" | 64 |
| 18 | John Glenn and Yuri Gagarin | 67 |
| 19 | Alan, age 13 | 70 |
| 20 | Winning all the math awards, *The Roslyn News*, June 15, 1972 | 83 |
| 21 | Painting of Chevy Corvette and Americana, Dana Forrester watercolors | 85 |
| 22 | Article about the JFK assassination in MIT newspaper *Thursday*, November 29, 1973 | 99 |
| 23 | Marc Silver, my lifelong best friend | 102 |
| 24 | Albert Camus | 109 |
| 25 | Alan, age 19 | 112 |
| 26 | Jean Baudrillard and Guy Debord | 120 |
| 27 | U.S.S.R. Pavilion and French Pavilion, Montreal Expo 67 | 122 |
| 28 | Mural in the Bologna university zone from the student movement of 1977 | 136 |
| 29 | The DAMS dragon from the 1977 movement | 139 |
| 30 | Umberto Eco at DAMS during the Bologna student uprising of 1977 | 141 |
| 31 | October 1978 DAMS city bus activist art project, ironic gift to the city of Bologna | 145 |
| 32 | Fremont Street, Las Vegas | 148 |
| 33 | Statement of Philosophy of the "Free Association" alternative university | 161 |
| 34 | Portmeirion, Wales, UK, the filming location for The Village in the TV show *The Prisoner* | 174 |
| 35 | Roulette scene in the episode "A. B. and C" of *The Prisoner* | 178 |
| 36 | Alan, age 22 | 181 |
| 37 | Casino Express Boat to the Venice Lido casino | 190 |
| 38 | The closed and boarded-up Lido Casino, many years later | 194 |
| 39 | Near the "Crazy Bar," I hallucinated seeing the wormhole deep in the water of the canal | 198 |

| | | |
|---|---|---|
| 40 | Alan, age 24 | 201 |
| 41 | Letter of Recommendation from political philosopher Claude Lefort | 212 |
| 42 | Donald Trump loves McDonald's fast food and gambling casinos | 218 |
| 43 | *Telos* editor-in-chief Paul Piccone's nasty letter to me about my work, May 1985 | 224 |
| 44 | January 1984 Super Bowl TV commercial introducing the Apple Macintosh | 237 |
| 45 | Letter of Recommendation from Larry of Neuberger & Berman | 244 |
| 46 | C code that implements an algorithm of Euclid | 245 |
| 47 | C++ code that creates an object instantiated from a class and accesses its attributes | 247 |
| 48 | Code from the original Microsoft Windows Software Development Kit | 248 |
| 49 | C code for FTP file transfer using TCP/IP Internet Sockets | 249 |
| 50 | Letter of Recommendation from Steven of Neuberger & Berman | 252 |
| 51 | New York-New York Hotel and Casino, Las Vegas | 258 |
| 52 | Bell Tower of St. Mark's Church in Venice and in Las Vegas | 258 |
| 53 | Rialto Bridge and Rialto Deli at Venetian Resort Hotel Casino, Las Vegas | 259 |
| 54 | Whiskey Pete's Casino, Primm, Nevada | 264 |

# FOREWORD

This memoir or auto-socio-biography is about the relationship between America and Europe. I left the USA several decades ago. I was disillusioned with my country's rightward political and cultural turn. Yet I love my first "homeland" in many ways. I sought an existential and philosophical understanding of what I glimpsed as a possible secret connection between these two "continents" or "halves" of Western civilization, democracy, and capitalism.

I tell the story of my childhood and teenage years on Long Island. I write about my father, the son of Jewish Eastern European immigrants, who enlisted in the U.S. Army at age seventeen and was in combat in Belgium and France during the Second World War. I describe my experience of skipping ahead two years in school when I was seven and its consequences for my life and my education at MIT and Cornell University. I divulge the adventures I lived during my first four years traveling around Europe with almost no money and the strange and harrowing jobs I had in New York City. I departed to Europe for good at age thirty-five, when the chronology of the memoir ends.

After graduating with my B.A. from Cornell at age twenty, I became a traveler and a gambler. My experiences – both outside and inside casinos – in

## FOREWORD

Venice, Italy, and Las Vegas, Nevada, USA, crystallized symbolically the connection between America and Europe I sought. Venice and Las Vegas intertwined with one another.

There is a casino in Las Vegas where Venice is copied and simulated. It is the Venetian Resort Hotel Casino (renamed in 2021 to The Venetian). You can ride on a canal in a gondola steered by a gondolier wearing a picturesque costume. There are large-sized replicas of the Bell Tower or *Campanile* of Saint Mark's Square and the Rialto Bridge. There are ceiling paintings in the Renaissance style. The overall look of the property is Italian Gothic architecture.

I searched for authentic dialogue between Europe and America through the exemplary case of Venice and Las Vegas.

Venice is the ultimate European city, a living metaphor for all of Europe. It is one of the most beautiful cities in the world and a UNESCO World Heritage Site. Venice is the pinnacle and emblem of the most profound European treasures: history, culture, art, carnival, painting, architecture, and opera music. It was one of the principal centers of the Italian Renaissance.

Venetian Resort Hotel Casino, Las Vegas, Nevada, USA (Photo by Alan N. Shapiro)

Las Vegas is the unofficial capital of America's consumerist and spectacular self-image. It is the Entertainment Capital of the World, known for nightlife, luxury, shopping, and fun-intensive vacations. Dubbed also Sin City, Las Vegas is the icon of everything excessive and gaudy about America. Inexpensive, lavish hotel suites and plentiful buffet food are on permanent offer to seduce visitors into losing money at gambling.

In my biography, the two cities became interlinked and even inverted. I played roulette and blackjack at the Casinò Municipale di Venezia, near the Lido strip of beaches. I drank Italian wine and cappuccino and contemplated classical fountains and sculptures in Las Vegas.

Ca' d'Oro Palace on the Grand Canal in Venice (Photo by Didier Descouens, Wikipedia Creative Commons BY-SA 4.0 License) and Venetian Resort Hotel Casino, Las Vegas (Photo by Alan N. Shapiro)

By trying to reproduce something perfectly, you obliterate the original to create your artifact. You make simulacra that refer only to themselves. The copy and the original both become simulations. The wall between "reality" and its images tends to disappear. Making a replica at the Las Vegas Venetian Hotel of Venice's walkways, footbridges, canals, gondolas, and artworks transforms Venice's actual "endangered city" (rising water levels caused by climate change threaten its future) into an *artificial paradise or paradoxical museum of itself.*

What is a ceiling painting? In the 1965 film *The Agony and the Ecstasy*, Pope Julius II (played by Rex Harrison) asks Michelangelo (played by Charlton

Venetian Resort Hotel Casino, Las Vegas (Photo by Alan N. Shapiro)

Heston): "When will there be an end?" To which Michelangelo replies: "When I'm finished." A ceiling painting takes forever because it prepares itself to have many layers of meaning and be open to nearly infinite interpretations. However, there is a limit to the possible readings. At some point, they reach a limit. It is the arrival of a wise owl (the Hegelian Owl of Minerva), as in the scene in *Blade Runner* when Deckard interrogates Rachael. There comes a moment when a grasp of the material at hand is recognized and captured in the finishing touch.

I pursued the unity between Venice and Las Vegas, seeking the singularity that connects those two cities. This culminated in a hallucinogenic experience: sighting an astrophysical "extreme exotic phenomenon" of a *wormhole* that links two remote points in spacetime. I believed I saw the science fictional wormhole deep in the water in a canal in Venice. Moments later, I was jolted to the awareness that it was not "really" there, or at least the verifiable status of its existence was a question mark.

# THE BEGINNING

When I was six, my mother knew she wanted me to become a top natural scientist, mathematician, or engineer. Having tested at a very high IQ, my life's trajectory was decided in advance. Harnessing that innate cognitive ability, I would make a lot of money like my father, the successful civil engineer who designed the structural integrity of skyscrapers. The plan was confirmed throughout my childhood and teenage years when I won many math competitions.

Much to my parents' delight, I skipped two grades in school, a marker of my so-called precocious genius. In my personal experience and inner life, a long period of social exclusion and quiet desperation began. I lay awake at night worrying about getting bullied the next day for being young-looking and a foot shorter than my classmates, not being able to compete fairly at sports, not having more than one or two friends, and the embarrassment of not knowing any girls to invite to my *Bar Mitzvah* reception. I was an outcast, an outsider, a subaltern.

When I was eighteen, I knew I wanted to be a philosopher, a thinker, and a writer. I toyed with the idea of becoming a professor of history or literature. Having become a political leftist as a consequence of the unjust Vietnam War and racism against African Americans and other minorities, I thought and felt that America, frankly, was a fucked up place. I wanted to get out, to leave America. This was one motivation for my wish to resettle in Europe.

What followed was a wild, diversified journey, influenced by films and social movements, chance encounters while wandering Europe and America, the need for money to survive, my dream of utopian social change, and my passion for some thinkers and writers.

I temporarily lent myself to many different careers and identities. I tried out potential lives. I abandoned those lives, put them aside to come back to later, or integrated them into the long-term invention of myself. What prospective lives did I try? What did I miss out on or not? What is the meaning (or its absence) of what I lived and the lessons of who I have become?

My life took me to participating by chance in a student uprising in Italy and "down-and-out" (George Orwell) jobs in New York City.[3] I battled with cockroaches in the studio apartments of slumlords, became Italian for a while, experienced windfall profits at gambling, followed lovers around, and aborted my academic career, about which I felt deep ambivalence.

I retell my story to shed some light on the socio-cultural circumstances of the historical era – 1960s to 1980s – of my childhood and youth on Long Island and in New York City, other places in America, and the early days of my European experiences.

Composing this memoir is an act of forgiveness and gratitude. It is an endeavor to upgrade or transform my memories into art – via the noble and healing act of writing.

# BOOKS BY ALAN N. SHAPIRO

*Star Trek: Technologies of Disappearance* (2004)

*The Technological Herbarium* (editor and translator, 2010)

*The Software of the Future* (2014)

*Transdisciplinary Design* (editor and contributor, 2017)

*Decoding Digital Culture with Science Fiction: Hyper-Modernism, Hyperreality, and Posthumanism* (2024)[2]

I thank Claudia Baniahmad, Bette Adriaanse, Fred R. Shapiro, Bernie Tuchman, Steve Valk, Jerry Rubin, William Miller, Marc Silver, Spiros Makris, Mike Gane, and Douglas Kellner for reading earlier versions of this manuscript and providing helpful feedback and advice.

CHAPTER 1

# When I Was Twenty

"I was twenty. I will let no one say it is the best time of life."

This opening sentence of the French writer Paul Nizan's travel memoir *Aden Arabie*, published in 1931, has always impressed me for its parallelism with what I lived.[4] Or maybe it was the opposite. Perhaps it was the best time of my life, and I did not know it. I wanted more from life than was my fair share. What I lived, what I was put through, could have killed me. That is the story I will tell in these pages. But since I survived and "made lemonade" from the lemons, as the psychotherapeutic saying goes, it is also a story of the best of all possible worlds.

At the time, I loved that sentence by Paul Nizan. Do I regret having been a rebel in my youth? I was a rebel all my life, yet not a revolutionary. The other students in my political science classes loved the anti-capitalist, anti-imperialist icons Che Guevara, Frantz Fanon, and Malcolm X. I think "ends justify the means" violence is unacceptable. State socialism from above is just as bad as, or worse than, the exploitative capitalist economic system that we have.

Others from my graduating high school class knew exactly what they would do with their lives. And they went on and did it. There was no doubt for them. Or was there? Andy was creative and spent some years as a visual artist. He then became a dentist, like his father. Mark took over his father's plumbing and heating contracting business in Lower Manhattan. Rick inherited his father's mechanical engineering firm, which provides building services in fire protection, heating, and ventilation. Kenny attended law school and became an accomplished legal expert in employee benefits and executive compensation packages.

Other friends from my college years who were more intellectual than my high school cohorts went on to graduate school in the social sciences or humanities. They got their doctoral degrees and continued a smooth path to academic tenured positions. I rejected everything that my society and my parents wanted me to be.

# CHAPTER 1

I had no idea what I would do with my life. I decided to leave New York, leave America, and go to Europe.

Paul Nizan was a mid-twentieth-century French Communist writer. He was a novelist, a political essayist, and a childhood friend of the existentialist philosopher Jean-Paul Sartre. Nizan was killed in the Battle of Dunkirk in 1940, in the early days of World War II.

In his memoir *Aden Arabie*, Nizan writes about being lost at the threshold of adulthood. Submerged in a personal and psychological crisis, he interrupted his studies at the elite École Normale Supérieure at age twenty to spend six months in the faraway land of what is today called Yemen, at the southern tip of the Arabian Peninsula. He was employed as a tutor to the son of a wealthy businessman. While in exile from the French educational establishment, Nizan sorted himself out. He came to an understanding of who he was. Could I follow in his footsteps?

## Cadillac to JFK Airport

It was two months after my twentieth birthday. My father drove me in his Cadillac Coupe De Ville from my parents' upper-middle-class split-level house in suburban Roslyn, Long Island, to JFK Airport in Queens. I was going to get on my Icelandic Airlines flight to Luxembourg, with a stopover in Reykjavik, to begin my first attempt to create a life for myself on the other side of the pond. I had an open return ticket that cost four hundred dollars – a lot of dough – and was good for one year. I just graduated with my bachelor's degree in history and literature from Cornell University at the absurdly precocious age of twenty. I was two years younger than my classmates because I skipped two grades in school when I was very young. It was the Cold War and the Space Age. America needed little math and science geniuses. My parents did not consider the psychological consequences of "skipping."

I embarked on my anticipated misadventures, hippie wanderings, and longshot try at settling somewhere in Europe. I had a small backpack. It had very few items inside: one extra pair of jeans, two shirts and a sweater, many underpants from Macy's Department Store following the advice of my mother, six pairs of socks, a few novels, a book of anarcho-Marxist political theory, a spiral notebook, two felt tip pens and a toothbrush. My dental care was poor during that period of my life. I went weeks without brushing my teeth. I would later need a full round of periodontal treatment to heal my gums and coax my mouth back to health.

I wanted to travel as lightly as possible. I had a cotton-filled blue sleeping bag. When rolled up, it fitted into a weatherproof green nylon cover. I had no idea where I would be sleeping – in youth hostels and cheap hotels, in train station waiting rooms, and on beaches. Living room floors waited for me, offered by hospitable people I would meet on the road.

I left New York with twelve hundred dollars in traveler's cheques in my pocket. Traveler's cheques were a safe way to carry money. If they got lost or stolen, the bank would reimburse you. You wrote down the serial numbers and kept them in a secure place. Twelve hundred was not much, but it was not nothing. My rough plan was to spend one hundred fifty bucks a month, mostly on food, to stay alive. There would be almost no funds for anything else. At that rate, the money would last about eight months. Some of it was from *bar mitzvah* presents. When I was thirteen years old, I endured that ritual ceremony of Judaism, memorizing in Hebrew portions of the *haftarah* from the Old Testament, singing them loud in a cadenced chant on a Saturday morning in April at the podium, on the stage of the synagogue. Some of my travel money was from summer jobs delivering packages and painting houses.

As a college student, I subsisted on nearly nothing and gave hardly any thought to money. I lived in the bubble of the university campus. My parents paid my tuition, which in the 1970s was only about three thousand dollars a year for an Ivy League education. I looked forward desperately and hopefully to the end of all the disciplined studying that had consumed my childhood and teenage years. All that I had in my mind was my fantasy of limitless freedom. I wanted to neither study nor have a job, to be away from all institutions with their bureaucratic procedures. I wanted to ramble – like the song "Ramblin' Man" (1973) by the Allman Brothers or like Jack Kerouac in his novel *On the Road*.[5] On a Greyhound bus, on a jet plane, on a train southward through Italy, on the side of a highway with my thumb stuck out.

Disillusioned by the Vietnam War, inspired ideationally and in sensibility by 1960s counterculture, and having read many French and European poets, novelists, and social critics, I wanted to escape America.

My Dad and I sat in the front seat of the Cadillac. My father had very little understanding of why I was going to Europe. My parents were from the "silent generation." There was no communication. They did not talk with me about anything having to do with life. They wanted me to pursue a money-making profession. They actively opposed my choice to study the humanities. It was Saturday the third of July 1976, the day before the maximum mega-spectacle of America's bicentennial celebration. It would be two hundred years to the day since the signing of the Declaration of Independence. There would be fireworks everywhere, especially on TV and in all media. There would be endless self-praising self-congratulations about the greatness of "America." I intentionally bought my plane ticket to Europe for the day before the extravaganza. It was time to

CHAPTER 1

go. My father was happy to drive me to the airport in his air-conditioned car. It provided him with a couple of hours of respite from the henpecking demands of my mother. He balanced being king at work and valet at home.

It was hot and muggy in New York and on Long Island, afternoon temperatures above ninety degrees Fahrenheit. The daily "news from nowhere" was that North and South Vietnam were reunited into one country.[6] There was an attempted coup in Sudan. There was a deadly terrorist explosion in Argentina. The Supreme Court declared that the death penalty was OK.

"The Mets have a good team this year," said my Dad.

"You mean those lovable losers of baseball don't suck as usual," I said.

"They still have some of the spirit of 1969. The World Champion Miracle Mets. And the 1973 team that went to the World Series. Jerry Koosman is having a great year," he replied.

"Yes, he is. He is my favorite pitcher because he is left-handed like me. I remember when he was 11-2 [eleven wins and two losses] in the first half of his 1968 rookie year. That was before the Mets were any good. This current eight-game winning streak feels good. I wonder if they can keep it going. They've never won more than eleven in a row in their history."

"How are you going to follow the Mets in Europe?"

"I can read the *International Herald Tribune*. They have a good sports section. But they report on the game from the day before yesterday. They show the 'League Leaders,' but not all players' batting averages and stats, as the *New York Times* does on Sunday."

"Your mother and I went to Europe once. You should go to Venice and see the old Jewish ghetto. It's still well-preserved. Write us letters or send us a postcard."

"I will do all of that. Bye, Dad. I love you." I got out of the car, one leg at a time, grabbed my small backpack and sleeping bag, returning from the cocoon of Cadillac heaven into the dust and grime of the physical world.

I still had a few hours until my flight. I planned to sit in a lounge and read George Orwell's early novel *Keep the Aspidistra Flying*, which discusses the banality of English middle-class life.[7] I walked past a man in a sweater holding a little girl in his arms as they looked together through a window at the nose of a Jumbo Jet. My mind asked itself about the broader historical context of what brought me to where I stood now.

## From World War Two to Suburban Long Island

The classic era of white American prosperity followed the Second World War. It was the famous 1950s and early 1960s blast of consumerism, film-and-television media-and-image culture, advertising by hidden persuaders, and the happy heterosexual nuclear family. It was the Great Lost Time of which the Trumpian MAGA cultists still dream. Hollywood films such as *The Truman Show*, *Pleasantville*, and *Don't Worry Darling*, as well as TV series like *Mad Men*, obsessively and nostalgically depict those times, albeit in ironic mode.

New Yorkers who belonged to some ethnic groups and had upwardly mobile incomes moved outside the city to the spacious neighborhoods and towns of Long Island. The shiny new houses and opulent shopping centers were within reasonable commuting distance from middle-class jobs in Manhattan, endurably reachable by car or the Long Island Railroad train.

I was born in Queens Jewish Hospital, which was in Brooklyn, in 1956. It was the year that, in the baseball World Series, Don Larsen pitched his perfect game for the New York Yankees against the Brooklyn Dodgers. Future Hall of Fame catcher Yogi Berra leaped into Larsen's arms after the twenty-seventh out. "It ain't been done before, and it ain't been done since," Yogi, the sage philosopher of baseball and life, deadpanned.

No need to brush my teeth yet (Alan N. Shapiro private collection)

My father, Murray, was the son of Jewish immigrants who escaped pogroms in south Ukraine shortly after the 1917 Russian Revolution. My grandfather had fifteen brothers and sisters. Some died in childhood. Some went to the territory that would become Israel, some emigrated to Germany, and some to New York. My grandfather owned a small grocery store in Flatbush, Brooklyn. Dorothy Walton Reese, the wife of Dodgers baseball shortstop legend Harold "Pee Wee"

Reese, was a regular customer in his store. It was said that Eleanor Roosevelt also stopped by. There is no public record of her living in Brooklyn, but she gave an inaugural speech in November 1932 at the opening of the Sears Roebuck & Company Department Store in Flatbush. Perhaps my father's family lived in the Bronx before moving to Brooklyn. Maybe Mrs. Roosevelt shopped in an earlier location of my grandfather's grocery store.

My Dad was a Brooklyn Dodgers baseball fan. He remembered many details of the first game he ever attended at Ebbets Field – in 1937 at age eleven. The great Carl Hubbell pitching for the New York Giants. Van Lingle Mungo was on the mound for the Dodgers: the high-kicking righthanded desperado fireballer later immortalized in the 1969 novelty song (entitled "Van Lingle Mungo") by Dave Frishberg about mid-twentieth-century baseball players.

"Two and two, what'll he do?" asked the radio ad, etched in my father's memory. "Buy a Goldenberg's Peanut Chew."

My parents told me my father also "skipped" two years in the New York City public school system in the 1930s. My older brother Fred also skipped two years. Like me, they both graduated high school at age sixteen. Unlike me, they did not seem to mind.

Fred told me recently that he does not think about his childhood because he does not particularly like the person he was and prefers the person he is now as an adult. It is almost as if the "childhood Fred" is, for him, a different person. He says that he is not afraid of self-examination, but he thinks it is a better use of his time to live in the present.

Why my father "did not mind" skipping two grades in school is an enigma to me because he never talked about it. I felt too intimidated by the taboo in my family when speaking about personal things, and I never asked him questions directly. Superficially, one might say that achievement was his highest priority and that psychology did not exist for him.

My Dad enlisted in the U.S. Army at age seventeen to fight the Nazis. He was sent to Texas and Louisiana for basic training. He was in combat in Europe for six months as a Private First Class and anti-tank specialist with the First Infantry Division. He fought in the Battle of the Bulge (*la bataille des Ardennes*) in Belgium, where an estimated 185,000 soldiers were killed, wounded, or declared missing. A shell exploded a few feet from his head. He was thrown violently to the ground from his gunner's position and cracked his skull. After several days of unconsciousness, Murray woke up in the hospital, partly deaf with ringing in his ears for life. Combat was a brutal experience. "We didn't sleep day or night," he told me. "We slept whenever we could, in fits and starts. We were hungry, cold, and tired all the time."

Here is a letter, written in the European war zone, from my father to my grandparents:

Dear Folks,
    I've got a chance to write, so I hasten to do so as these chances seem to come less and less. I'm well and feeling fine. There is nothing unusual to write about. You see that I still have the same address. It stays the same even though I have now joined an outfit.
    I ran into a fellow I knew from City College. It was swell to talk about some of our experiences back home. Incidentally, Bob Schulman, whose address I sent you, is also a city boy. He attended school downtown and is taking up accounting.
    While in Belgium, I was invited to a home, and it was very nice to get into friendly, warm surroundings again. Of course, our family's home is always in my mind and heart, and you all are dearer to me than life itself. I'm afraid it has taken a war to show me how much you mean to me. I'm as fortunate as very few can be to have you home waiting for me. Believe me, whatever I do, wherever I go, you all are my inspiration and guidance.
    I pray God be with you always, as I know you pray for me. We are miles apart, but I feel we go through these times together, the four of us, and nothing, nothing at all, that may befall us can change our unity.

All my love from the bottom of my heart with all my heart,
Moishe

The American men and women who risked their lives in the Second World War fulfilled their fundamental moral duty. They fought an enemy, the "evil" of which was an unambiguous given. Courage and personal sacrifice were the routine conduct of this "band of brothers."[8]

The survivors came home to an America of unprecedented economic opportunity. The collective project of making the American dream into a "hyperreality" unfolded.

Hyperreality is the artificial staging of "reality" by the cultural models and codes that precede it.[9] Images become increasingly autonomous from the originals they were supposed to be merely copies of. Rhetorical discourse – what today is called "post-truth" – becomes increasingly independent from the facts and "references" it previously allegedly described.

Mass identity architecture designs simulation and "nowhere" spaces like Disneyland, shopping malls, airport lounges, and suburban housing developments. In hyperreality, simulation supersedes representation. In the post-World War Two era, media and consumer culture overtook America.

Thanks to the 1944 "GI Bill of Rights," affordable higher education and loans for new home construction were accessible to returning veterans. My Dad, Murray, was a whiz at math and logical thinking. He earned a degree in civil engineering from City College. He became a highly successful structural engineer in Manhattan, working on skyscrapers.

My Dad designed over one hundred significant high-rise office buildings in New York City and Boston. Some of his notable accomplishments include the Pan Am Building (now called Met Life), the General Motors Building, Citigroup Center, One Penn Plaza at Madison Square Garden/ Penn Station, 55 Water Street (at one time the largest office building by floor area in the world), the Art Deco-style tower at 1675 Broadway, and the John Hancock Tower in Boston.

In the early 1960s, my father worked out the safety logistics of the monumental Pan Am Building, constructed over the north shed of Grand Central Terminal and atop an eight-story granite base. He worked with famed Bauhaus architect Walter Gropius. Completed in 1963, the skyscraper, initially owned by America's foremost international airline, is the largest commercial office building in the world. It is nearly four hundred thousand square meters of rentable space. The vertical structure, covered on the outside with concrete panels, towers over the middle of Manhattan.

The edifice's octagonal shape and east-west perpendicularity to the buildings lining the north-south streets institute a singular communication with peers. The structure divides Park Avenue into north and south segments rendered invisible on the ground to each other. On the skyscraper's flat roof, the heliport of NY Airways offered a speedy transfer to LaGuardia, JFK, or Newark airports – a thrill ride to be discontinued in 1977 after a fatal helicopter crash. In John Carpenter's 1982 science fiction film *Escape From New York*, Snake Plissken (played by Kurt Russell) lands his one-man stealth plane perilously onto the flat top of one of the World Trade Center Twin Towers.

According to architectural critics like Ada Louise Huxtable and Jane Jacobs, the Pan Am Building brought congestion to the midtown East Side, blocked the magnificent Park Avenue vista, and shrouded the iconic masterpiece of the New York Central Building.[10] To the contrary, I see the Pan Am Building as an ironic "monster" architecture (in the sense theorized by Jean Nouvel and Colin Fournier) that "issues a challenge" to the urban space of New York City.[11] It is a *radical illusion* beyond its designers' officially lamented lack of aesthetic sensibility. The structure is dense but leaves space for movement. There is an elaborately engineered system of human circulation between places of business and the train terminal. The high-speed elevators and array of escalators are complemented by the passageways and tunnels leading to rail and subway service. Poised above the commuter stations, the ground floor features an open expanse leading to the ticketing promenade around the majestic analog clock.

Home on its upper floors to tiny windowless offices of currency changers and translation agencies, the Pan Am Building was symbolic of a cultural exchange of America with the world.

I am looking at the handwritten list that my father – when he was over seventy – drew up of all the buildings and constructions for which he did the principal structural engineering. He did many theaters in New York and several theaters in Israel. He did many high schools, university buildings, hospitals, churches, synagogues, Ferris wheels, hotels, bus terminals, public libraries, and department stores. He did the Time-Life Building in Chicago. He engineered many Philadelphia, Pittsburgh, Atlanta, Baltimore, and Washington, D.C. buildings. According to the list, he did twenty-two office buildings on Park Avenue in Manhattan, eleven on Third Avenue, five on Lexington Avenue, eleven on Madison Avenue, seven on Fifth Avenue, seven on the Avenue of the Americas (Sixth Avenue), nine on Broadway, and nineteen high-rise buildings in the downtown Wall Street area.

A handwritten list made by my father shortly before his death (Alan N. Shapiro private collection)

For thirty-five years, my father commuted five or six days a week – by bus and subway, later by car – between Long Island and midtown Manhattan. Cruising in his Cadillac, he knew all the secret shortcuts to avoid traffic. Take the Grand Central Parkway past the World's Fair Grounds and LaGuardia Airport (close to Shea Stadium, the home of our beloved baseball team, the Mets) to just before the Triboro (now called Robert F. Kennedy) Bridge. You turn left under the elevated subway line. You cut through backstreets.

CHAPTER 1

You pass what I called "Sneaker City" – where urban teenagers threw dozens of pairs of sneakers tied together with laces over telephone wires – down to the Fifty-Ninth Street Bridge. Those "shoe trees" or *shoefiti* challenged the ruling order of messages and meanings of the "plastic" consumer culture.

While driving on the Parkway, Dad could save three minutes by hot-rodding his car through the service station's exit and re-entry ramps just before LaGuardia airport while the cars on his left stood still.

For me, the Fifty-Ninth Street Bridge was immortalized in and by the final desperate car chase sequence in *Escape from New York*.

In the late 1950s on Long Island (the era when *Mad Men* takes place), agricultural tracts were *en masse* converted to housing developments. It was an immense transformation.

My parents, Murray and Florence Shapiro, in the early 1950s
(Alan N. Shapiro private collection)

In my earliest memory, my mother is buying fresh corn on the cob. She shops at the farm stand five minutes from our newly erected tri-level house with three staggered levels and two staircases on Gordon Drive. When I was two years old, my family moved from an apartment in Forest Hills, Queens – about which I have no recollection – to a middle-class home with a backyard in Williston Park on the North Shore of Nassau County. The architecture of the house was *postmodern*. It was one of a "mass identity" series of twenty-six houses that all looked exactly alike, evenly spaced in small property plots on two blocks of street, fabricated or generated from the same cookie-cutter model. It was the same principle as manufacturing Snickers Bars or Twinkies, products coming off the assembly line, or an Andy Warhol lithograph painting of Campbell's soup cans.

The model house of our cookie-cutter house, viewed from the backyard, Williston Park, LI, NY, the year 1958 (Alan N. Shapiro private collection)

I left Long Island at age sixteen and never returned. My understanding of the cultural characteristics of Long Island was and remains grounded in the consciousness of a child. I have never tried, nor wanted, to change that awareness. After the horror of World War Two and the six million Jews killed in the Holocaust, everything on Long Island in my perception and experience was, as Madonna sang in her song "Like a Virgin" (1984), shiny and new:

· The young President John F. Kennedy
· The Mercury, Gemini, and Apollo space programs

CHAPTER 1

- The local shopping center close to our house, featuring the "toy store" and the bakery with amazing chocolate cookies
- Candy: Peppermint Patties, PEZ dispensers, Pixie Stix candy straws, wax candy bottles with sugary juice
- Baseball cards in packs
- Comic books: DC Superheroes, Classics Illustrated
- Dunkin' Donuts
- The bowling alley
- The movie theater
- Hamburger, French fries, and a Coke
- A slice of pizza
- TV Guide magazine with the Fall preview of the new shows
- The first McDonald's and Kentucky Fried Chicken
- The Hostess series of cake products: Twinkies, Snowballs, and Yodels - exemplifying the "simulation of difference."

Everyone on Long Island had a car, and some families had two cars. Cars were symbols of social status and economic wealth. They were also an absolute necessity for everyday life's logistical survival in suburban America.

Not everything on Long Island was new. When I was fourteen, during the summer between the tenth and eleventh grades, we moved from Williston Park to a larger, more expensive, and more valuable house in Roslyn. There were old historic structures and houses in Roslyn Village – the Paper Mill (a simulation replica of the original Mill), the Van Nostrand-Starkins House that was constructed in the seventeenth century, many nineteenth-century pre-Civil War houses along Main Street, and the beautiful Ellen E. Ward Memorial Clock Tower.

I see the Roslyn Clock Tower in my mind when I read the Czech-Jewish writer Franz Kafka's famous parable "Give It Up!":

> It was very early in the morning, the streets clean and deserted. I was on my way to the station. As I compared the tower clock with my watch, I realized it was much later than I had thought and that I had to hurry; the shock of this discovery made me feel uncertain of the way. I wasn't very well acquainted with the town.
>
> Fortunately, a policeman was at hand. I ran to him and breathlessly asked him the way. He smiled and said:
>
> "You're asking me the way?"
>
> "Yes," I said, "since I can't find it myself."
>
> "Give it up! Give it up!" said he, and turned with a sudden jerk, like someone who wants to be alone with his laughter.[12]

In the town of my early childhood, Williston Park, there were old houses on Lafayette Street just beyond our two blocks of "new" serially duplicated dwellings. Those houses did not have the same bright outdoor paint colors as ours. Each domicile had its singular architecture. The people who lived there were mysterious and terrifying. They were not to be trusted. I was afraid to go to those houses for "Trick or Treat" on Halloween. I stayed away from them. They might have given me poison or a sharp utensil hidden inside the candy.

## Between the Cold War and Baseball

The crises of the Information Age are rooted in the Space Age. After the Soviet Union launched the Sputnik satellite into orbit in 1957, America responded – inspired by fear and panic – with its national project of developing a technocratic state that took direct responsibility for science and technology. Since the 1960s, we collectively exist in nonstop technological developments – first in outer space, then in cyberspace, and soon in the digital media technologies of the so-called Fourth Industrial Revolution: Virtual Reality, Augmented Reality, Artificial Intelligence algorithms, and the Brain Computer Interface – that bring along with them a permanent cultural crisis. Only a lucid utopian "think tank" vision in theory and practice of how these technologies could be designed and implemented alternatively, fully independent from the context of "surveillance or platform capitalism," could steer us out of the crisis.[13]

America was looking for techno-scientists from the ranks of the children of the middle class. It was 1962. I was six years old and tested at an I.Q. of 165. I was identified as a one in ten thousand "high intelligence" and sent to a special "school for gifted children." There, I completed six years of elementary school in four years. I re-entered the public school in the seventh grade. I was ten years old. My classmates were twelve. This "being two years younger" made my adolescence socially miserable. It took me a very long time to recover psychologically from and to overcome that. I became a component of a system seeking out future technologists and scientists. No consideration was given to developing my social and emotional intelligence or interpersonal well-being. The Cold War, the Space Race, and the politics of technoscience and early informatics determined my fate in life.

Behind our house, we had what appeared to me as a child to be an extensive backyard. It was about thirty by twenty meters (about one hundred by sixty-five feet). Compared to the national average, this was rather on the small side. Every house on the block had the same-sized backyard. That green space was my world, my lonely cosmos. The grass was long, unruly, and uncut. There were swarms

CHAPTER 1

of honeybees who helicoptered in and around the grass in the summer. They flitted among the daffodils. I was a very gentle boy, but what sadistic streak I did have expressed itself in my stepping on honeybees and grinding them into the ground to their death. It was a risky business. If you missed with your sneaker, you could get stung by a bee gone wild. I wonder why I felt the drive to stomp on them so forcefully. I had repressed anger.

I played many solitaire sports, embellishing my ritual physical movements as imaginary matches in the famous American professional leagues fantasized with my mind. I staged elaborate NFL-AFL football games where I was both the quarterback and the receiver, throwing the prolate spheroid ball in a high arc as a long forward pass, then running to catch it or not. An interception by a defender was also possible.

I kicked my soccer ball between the imaginary goalposts of two bushes, slamming into the wooden fence of the neighboring property. Sometimes, this caused damage and a harsh reproach as the woman next door came out screaming.

I played golf with a two-way putter. I still have that battered golf club in my possession many decades later. Fred and I dug holes in the ground with a spoon to set up the golf course. We dug a broad and deep hole that we hoped would reach China.

We played ping-pong in our unfinished basement. Table tennis became my best sport.

Baseball was my primary sport and passion. I watched the Yankees in our "recreation room" on our black-and-white TV. My first dream profession was to be a baseball announcer. I did not hear about the Mets until 1966. I became a Mets fan in 1968, inspired by Koosman. I listened constantly to broadcasts of Yankee games on my little transistor radio. There were many afternoon games and Sunday doubleheaders. At night, I took the radio to bed with me. I listened secretly with an earplug, hoping that I would not get caught by my mother.

Game Seven of the 1964 World Series between the New York Yankees and the St. Louis Cardinals. On a cloudy afternoon after school, I am alone in the backyard, listening to Joe Garagiola describing the game. The Yankees hit two solo home runs in the top of the ninth inning. They still trail the Cardinals by two runs. There are two outs. The Yankees' hopes are hanging by a thread.

> Bobby Richardson is up there with two outs. Seven to five. He's got to get on, get the big bomber Maris up there with the tying run. Bob Gibson is ready and delivers. It's low. Ball one. One ball, no strikes. Roger Maris, kneeling in the on-deck circle with that big bat in his hand, waiting to get up there as Richardson takes a strike. One ball, one strike.
>
> This ballpark just bulging with every pitch. One ball, one strike. Richardson waits. Gibson delivers. Swung on, popped up. Maxwell is at second base, calling for it, and

34

makes the catch! The Cardinals win it. And this ballpark is complete bedlam. The final score is St. Louis 7, New York 5. And in a moment, we will review the game's highlights for you.

The Yankees lose the World Series. Deep depression sets in.

I had a hard ball and a pitcher's glove. I became skilled at throwing the ball vertically and catching the pop flies as they came down. I could throw the ball sky high and far ahead, then run to get under it and sometimes even dive. I had a rubber ball that I threw against the brick wall of the back of our house as part of an elaborate baseball simulation game.

I played that simulation game for years, although my mother tried to stop me. The brick wall was between a window and the sliding glass door. I was always at risk of breaking one of them, but I never did. There were four bricks in a diamond pattern that were the strike zone. Anything outside the strike zone was a ball unless it was a hit. My ambition was to be a pitcher, so I threw the rubber ball very hard. There were two bricks above the strike zone that were the hit area. Any pitch that impacted either of those bricks meant contact by the batter and a live ball coming back at me that I had to field. If I snared it with my glove either in the air or after it bounced, it was either a fly out or a ground out. If it got past me, it was a foul ball or a single, double, triple, or home run, depending on where the ball stopped in defined territories.

Behind our backyard was the vast and savage baseball field. There were additional flat and rolling "fields of dreams." They belonged to the local elementary school.

That single-story school I saw on the horizon was the elusive foreign promised land to me, the distant site of my estrangement. I attended that school for one year in kindergarten. My parents yanked me out and sent me on a daily bus ride to the private school for gifted children. On countless occasions, I watched the "normal" children in their recess and lunch period mingling socially and playing games on the blacktop recreation and basketball surface in the distance.

All the other houses in the uniform row on our side of the street of Gordon Drive had high wooden fences that a child could only climb over with great effort. This kept the schoolkids from "cutting through" your backyard as their shortcut to the shopping plaza and beyond. My mother was an oil painter. She wanted an unobstructed view of the unbounded open space of the crude baseball field. She installed a rustic ranch fence with two low-height horizontal crimped rails in each paling section. This flimsy fence made us the object of ridicule of the local youth. I would be trying to enjoy my solitary life in the backyard while older boys trampled through with smirks on their faces.

CHAPTER 1

Florence Morrison (my mother), Connecting Fences, 1960 (Alan N. Shapiro private collection)

One of my motivations for writing this auto-socio-biography memoir is to express my alienation from the society where I was born. Marxist and leftist political philosophies place their emphasis on the subaltern status of groups of people designated as structurally disadvantaged by and in the "order of things" – the workers (the proletariat); the Third World colonized; people of non-white skin; women; gays, lesbians, and transgenders. This is called identity politics. According to the conceptual system underlying those classifications, I would have nothing to say as a witness to social oppression. I am white, male, Western, heterosexual, and from the middle class. I am Jewish – not sure where that fits into the schema.

My only legitimate link to fighting for social change would be as the concerned intellectual who looks at the exploitation of the worker by the capitalist or of the colonized by the colonizer and feels morally compelled (from the outside) to do something about those situations of injustice. My claim, on the contrary, is that the pivotal problem of our society is not the oppression of groups but rather the alienation of the individual. Through telling my story, I stake my claim to the right to speak about what is wrong with our society. I recount my tales. How did I survive my alienation? Where did my journey go? What is the lesson from my quest for meaning? Writing is the triumph over what I was put through.

CHAPTER 2

# The Scene of the Crime

## Nursery School, Kindergarten, First Grade

At age four, I attended the local nursery school on Hillside Avenue between Temple Emanuel synagogue, the skating rink, and the giant food center. I would later attend religious services on Friday evenings and Saturday mornings at the Jewish place of worship. Skateland ice rink was the practice facility of the New York Rangers National Hockey League team.

At the Big Apple supermarket, I bought powdered sugar donuts and inserted a nickel into a bubble gum machine to get a pack of five "Mars Attacks" cards. Starting at Card Number 46 of the 55-card series, the Earthlings launch their counterattack against the brutal and hideous Martian would-be colonizers. After their "Blast Off for Mars," the heroic human volunteers bomb the Red Planet and leave the Martian cities in ruins. It was an echo of H.G. Wells' science fiction novel *The War of the Worlds* and a metaphor for America's paranoid fear of an imminent assault by the international Communists.[14]

From the back side of the card:

> Military units worldwide reported to their nearest rocket bases, quickly set up by the newly formed 'Space Committee.' Governments from every country on Earth worked hand in hand to fight the menace from Mars. Tanks, guns, rocket planes, and soldiers were loaded into spaceships, ready to continue the vicious war with the Martians on Mars. Men aged 16 to 45 were given quick physical examinations and enlisted into the Earth Army. Volunteers flooded the recruiting centers, as brave citizens showed they were eager to fight the worst peril ever threatening civilization. (Len Brown inventor, Topps Company, Inc., 1962)

The preschool romper room was in a small building more suitable for a bakery or a take-out restaurant. There were two groups of about twenty kids each. To temper the chaos, the other children and I took many mid-morning naps on the broadloom carpet. We listened to a scratchy vinyl record of the Dance of the Sugar Plum Fairy from Tchaikovsky's Nutcracker Suite.

The following year (age five), I was in kindergarten at the public Center Street Elementary School behind the baseball field, behind the backyard of our house. In kindergarten, I danced "The Twist" (1960) to Chubby Checkers' popular song promoted by the music industry. I made gyrating movements with my body (resembling the solo Twist dance) while keeping a Wham-O plastic toy hula hoop in place. I twirled the bright red hula hoop around my waist for as long as possible before eventually losing control, and gravity pulled it to the ground.

Since I tested very high IQ, my mother decided I was bored in kindergarten. She thought I would continue to be bored with the learning curriculum the following year if I stayed at the level and in the company of other pupils of my age group in the first grade. She met with the school principal and tried to persuade him to let me advance or "skip" immediately into the second grade that coming September. He refused her request. The educational administrator believed that "skipping" was a bad idea. He thought that the scholastic and "achievement" dimensions of my existence were not everything. His view was that my being in a grade where I was younger than my classmates would be socially, psychologically, and emotionally damaging to me.

My mother left the meeting in anger. She withdrew me from the public school and enrolled me in the private Sands Point Country Day School for Gifted Children.

The Sands Point School was housed in a sumptuous mansion on a sprawling real estate property adjacent to the rough waters of the Long Island Sound estuary at Hempstead Harbor. The regal manor house on Elm Court was built with "old money" wealth around 1921, at the dawn of the Roaring Twenties. The famous architect Egerton Swartwout designed the stately home of twenty-two rooms, built with concrete and stucco, for the shipping magnate Edgar F. Luckenbach, President of the Luckenbach Steamship Company, whose wealth was inherited.

The estate became renowned for its staging of the Sands Point Horse Show. Luckenbach's third wife, Andrea Fenwick Luckenbach, was a dedicated breeder of exhibition horses. The celebrated French landscape creator Jacques Gréber (planner of the Ben Franklin Parkway in Philadelphia) added sumptuous formal gardens, lawns, orchards, greenhouses, elaborate fountains, and an intricate hedge maze to the idyllic environs around the villa. The imposing residence-converted-to-an-elementary-school stood proudly in the center of 140 acres (57 hectares) of land, forest, and beachfront.

The Sands Point Gold Coast is the geographical locale corresponding to the fictionalized East Egg, where F. Scott Fitzgerald's iconic novel *The Great Gatsby* partly takes place.[15]

Sands Point Country Day School mansion at Elm Court
(Photo by Academia45, Wikipedia Creative Commons BY-SA 4.0 License)

My successful structural engineer father paid tuition money to the private school. The owners agreed, in advance, to let me "skip" my way through grade school as fast as my "big brain" capabilities would carry me.

Many times, up until my mother's death at age ninety-one, I asked her to acknowledge that the skipping was a big mistake that harmed me or at least caused me suffering and made my life difficult. She never budged in her position. Every time I brought up the subject, my mother repeated insistently her stance that I would have become a "discipline problem" if I had been left at age six in the original grade track of children my age. My mother never apologized nor developed self-criticism about the action that deeply affected and shaped my life.

Ironically, the Sands Point School was, in fact, a haven for pupils with "discipline problems." Many articles about the school appeared in prestigious New York newspapers and national magazines in the 1960s, praising it as an advanced, enlightened project alleged to be vitally important to America's future. The school was showered with accolades. It could model how gifted children of high intelligence and extraordinary creativity could be pedagogically nurtured. The school came to the attention of national TV network news programs, the U.S. Congress, Senator Robert F. Kennedy, and Nobel Peace Prize winner Ralph Bunche.

The dirty bottom-line secret, however, was that the business enterprise that was the school was keenly in need of money. It was willing to take almost any pupil whose parents could afford the tuition. A significant minority of the kids were "gifted" with high "measurable" IQ intelligence. But many were problem cases. They were expelled from their public schools for engaging in various mischiefs. They needed to be accepted somewhere. Perhaps the parents wanted to believe that the juvenile delinquency of their son or daughter was a spot-on marker of great genius. There was a third large category of ordinary kids with average intelligence whose parents had money and wanted to feel their child was unique. Everything in America was the reign of the image, the society of the spectacle. Sands Point was the simulacrum of being "gifted."

I woke up early five days a week for the bus ride to the private school. I was required to be at the appointed spot on the sidewalk of the thoroughfare of Herricks Road at the designated time of 8 AM. Gordon Drive was connected to Herricks Road at both ends of our short C-shaped lane. The school van picked me up to begin the seemingly interminable drive to the Sands Point mansion. The bus made several further stops to collect other children. The trip could take one hour, depending on the heaviness of the "rush hour" traffic and the weather. It could be miserably cold and snowy in the depths of winter. When we finally arrived, I was exhausted and freezing and felt like my energy for the day was already spent.

Now I close my eyes and remember. The concept of memory is one-sided. It is valid but not sufficient. Each lived moment gets circumscribed by the past tense but is also vital independently of time.

The bus turns off Middle Neck Road onto the tight, secluded pathway of Elm Court. The speed limit is nearly zero. The vehicle proceeds deliberately and cautiously. The driver is attentively looking for detached tree branches which may have fallen during the night's storm. At the end of the pastoral road is the circular driveway in front of the school entrance.

The manually operated frame folding door of the bus-sized van swings open. I descend to the ground. I take a few steps up to the portico gateway that shelters those who arrive from inclement weather. I observe its roof structure supported by elegantly chiseled columns.

I glance behind the grandiose structure at the Jacques Gréber-designed complex labyrinth with its manicured evergreen shrubbery. I often intentionally get lost inside that maze during recreational recess or lunch hour. Then I find my way out like ancient Theseus, searching for Cretan Princess Ariadne's thread to escape the menacing Minotaur.

It is just before 9 AM. I step into the vestibule. I walk through a wide corridor hallway. Essential rooms are on the ground floor: the headmaster's office, the music room, the science room, the library, the dining hall, the kitchen, and the walk-in closet for mimeographing.

My legs, however, guide me towards the "grand stair" – a broad winding staircase veering left that leads majestically up to the second floor. I ascend with my left hand on the metal handrail at the mahogany banister, espying the ornate chandelier light fixture overhead.

I reach the second floor. To my proximate left and a few meters away, and my immediate right, and in front of me beyond an open space, and to the distant left and on the far right, I see the four classrooms where my life's fate would soon play itself out as early in the metaphorical day as the rooster rises in the morning on the proverbial farm.

The four classrooms were converted from their bedrooms at the time of the Luckenbach family. I enter the room directly on my right. The other rooms are the homes of my future second, third, and fourth grades.

The first-grade chamber is mine at age six. It is about five meters by five meters. It has an attached bathroom. It has a sleeping porch enclosed by mesh screens. Tall, arched Palladian windows adorn the side opposite the door to the room.

There were fifteen kids in the class. The teacher was kind, young Mrs. Wagner, the school principal's daughter. She had brown hair and wore glasses. She and my mother taught me to read.

The other boys and I spent a lot of time tossing and flipping baseball cards. You stood at a certain distance from the wall. You and your opponents tossed your cards, one by one, as close as possible to the wall. The best result of the toss would be the card touching the wall with one edge while it stood up diagonally. The winner gathered up all the cards in play. In another variation of tossing, you tried to cover the card of an adversary with your card. Flipping was a heads or tails game. Your card landed face up (showing the image of the baseball player) or face down (showing the back of the card with the player's career statistics). The second flipper tries to match the outcome of heads or tails.

We already had French classes in the first grade, a simulacrum meant to signify our smartness.

## CHAPTER 2

Four first-grade pupils, including me, were chosen to learn the second-grade curriculum in reading, writing, science, and arithmetic. Mrs. Wagner tutored us part of the time separately from the others. We felt proud of our exceptional advancement through the educational system. It was understood that our task was to complete the first and second grades combined in one year. It was understood that, in the autumn, we would go to the third-grade class.

Lisa, Barbara, and Danny were the other three kids in the accelerated group. I spent a lot of time with Lisa and Barbara. Years later, the moments with them became like a lost paradise in my mind. I would not have a girlfriend again until age eighteen. I was in love with Lisa. But I was even more in love with Barbara and her blonde hair. I was especially enamored of Barbara because she had the same last name as me. It felt, in a way, like we were married.

Danny was the most popular kid. He was smart – but nowhere near as smart as me. He was extroverted, self-confident, and good-looking. He was cool. But he was a small-time thug, a bully. He was dishonest and devious, not to be trusted. I was almost as popular as Danny. But I did not care much about being popular. Years later, to be liked and admired by the whole class became like another lost paradise. I would never experience that again in all my school years.

Danny declared to me and everyone else that I was his best friend. I would have made a good vice president or prince, serving under his rule as president or king. I humored him by saying I was, indeed, his best friend. In my mind, I disliked him.

One day, it was announced that there would be an election for class president and other officer positions. The nominations would happen today, and the voting would happen on Monday. But Danny was out sick that Friday. He was not there. No one nominated him. Despite his apparent popularity, he was forgotten. Out of sight, out of mind. Maybe secretly, no one liked him. They were just afraid of him. I thought of Danny. I considered if I should nominate him for president in his absence. But I decided against it. It was my revenge, my way of expressing that I did not like him. I was nominated for president, and so were two other boys.

At the beginning of the next week, Danny was healthy. He was there again, and he was furious. He was angry at me. Based on the rules, it was too late for him to be nominated for president. He launched the "Don't vote for Alan" campaign. All the kids – except for Lisa (my true friend) fell into line behind him. They made signs that said: "Don't vote for Alan." The negative posters were hung all over the classroom and the hall corridor.

"Don't vote for Alan! Don't vote for Alan!" they shouted.
I got two votes: Lisa and me. I felt terrible. Mrs. Wagner felt bad for me.
"What lesson did you learn from this experience?" she asked me.
"I learned that the world is full of strife," I replied.

I did not feel humiliated. I thought everything in the world was absurd. When you have heartfelt detachment, nothing humiliates you. But you can feel bad.

## Second Grade, Third Grade, Fourth Grade

At age seven, I returned to the mansion's second floor in September for my second year at the Sands Point School. Barbara and Danny were nowhere in sight, but Lisa was there. No one remembered that Lisa and I had completed the second-grade curriculum the previous year. It was a bureaucratic oversight. We were assigned to the second-grade class.

Now, I walk up the grand staircase steps, walking left instead of right. I sit at one of the wooden solo desks with its attached chair and rotating desktop. I wait for the right moment, then raise my hand.

"Yes, Alan?" says the female teacher.

"Lisa and I are not supposed to be in this class," I assert. "We are meant to be in the third grade. We did the second-grade curriculum last year combined with the first-grade one. We are supposed to skip to the third grade. It was agreed upon and promised."

The rest of the class looks around in befuddlement.

"I will discuss it with the principal, and we will see if that's true," replies the teacher.

My second grade lasted five days. I spent one week in that second-grade classroom with children my age for the last time – and it was boring. My mother was right about that.

On the Monday of the second week of the new school year, Lisa and I were told to go to the third-grade class.

I walk up the grand staircase steps and straight ahead. After fifteen steps, I enter the classroom on the right. It is another converted bedroom with another enclosed outdoor porch.

My assigned seat is next to a wall, at a double school desk, with a mate sitting with me. In this second year, Lisa and I are hardly on speaking terms anymore.

My third grade lasted two weeks. The other pupils were too slow for me, and the reading material was too easy. My comprehension was at a more advanced level. I already memorized the multiplication tables. I could write a short essay. I knew how to do research. I had some knowledge of American history. I had a sense of world geography. I knew the astronomy of the planets. I was aware of the different economic and governmental systems of various countries.

One Friday afternoon, at the end of the school day, Lisa and I were informed that we were directed to report to the fourth-grade class the following Monday. The principal called me into his office.

"You are too advanced for the third grade," he said. "It's not challenging for you. It's too monotonous. We believe in moving you ahead according to your abilities and beyond them. We believe in *stretching your mind*.[16] Jumping to the fourth grade is something to celebrate. It is a great achievement. You should be happy."

"Yes Sir. Yes, Mr. Director." *Oui, Monsieur le Directeur.* In my mind, I addressed him in French.

"Alan, we are a non-denominational school," he continued. "But all religions teach the belief in God. Their different versions of God are all essentially the same. You believe in God, don't you, Alan?"

"Yes, Mr. Director."

I could not follow what he was saying. I had no notion of God or religion.

It was probable that the principal discussed the promotion to the fourth grade with my parents, and even more probable was that they initiated the action. My "skipping twice" would now be completed.

"Now your mind will be appropriately intellectually stimulated," continued the principal. "We have considered maybe advancing you a third year, but that idea has been rejected, at least for now. Since you have a special aptitude for mathematics, you will go to the fifth-grade classroom for forty minutes each day – the math period."

"Yes, Mr. Director."

Suddenly, I am tossed into the cold, deep water of the fourth-grade class. I am plopped down into that setting. It is a crash landing. The other boys in the class are one foot (thirty centimeters) taller than me. They are giants, and I am small, a so-called "shrimp." I feel the difference between me and the others in a dramatic way. It was just the beginning of that. This, at age seven, is the determining moment of my life: the scene of the crime.

Now, I walk up the grand staircase steps and straight ahead. After walking fifteen steps, I enter the classroom on the left. It is another converted bedroom, yet smaller than the others. It is about 4 by 4 meters.

There are only nine kids in the fourth-grade class. I take my place on the long bench seat at a horizontally lengthy desk accommodating several pupils. Lisa was in the class, but we no longer spoke to each other.

I was in love with Anne, who sat at the desk immediately ahead of me. During the whole year, she never spoke more than a few words to me, nor I to her. I stared at the back of her medium-length blonde hair. Anne was six inches (fifteen centimeters) taller than me. She was the daughter of the school's French teacher. I thought Anne was French, but her family was from Poland. Her mother's French was probably not "native speaker French." Anne had a younger brother named Jacques, who was in the school in one of the lower grades. It was thought widely that he was a brat. At night, I fantasized I was the younger brother so I could feel close to her or that she would pay attention to me.

The kids in the fifth-grade math class were even taller. I sat in the back and kept quiet.

Mr. Hutner was the "homeroom" fourth-grade teacher. He also taught us American literature. He had a well-groomed mustache and was emotionally distant. He was philosophically opposed to the "skipping" policy of the school and resented my presence in the class. To demonstrate that I was really "not that smart," he assigned us to read the landmark early nineteenth-century work *The Last of the Mohicans* by James Fenimore Cooper.[17] Cooper's historical romance novel, with its unwieldy style and plodding plot, is barely comprehensible to sixteen-year-olds. I was seven. Yet I gave it a valiant try. The experience gave me a distaste for reading books, which lasted about ten years (although I always dutifully read what was assigned in school). This time, my mother was livid for the opposite reason. She came to the school and confronted Mr. Hutner for making things difficult for me and the others. After a few weeks, he switched us to reading *The Yearling* by Marjorie Kinnan Rawlings, a novel about poverty and the love of a pet.[18] That was followed, unsurprisingly, by Mark Twain's *The Adventures of Tom Sawyer* and *Adventures of Huckleberry Finn*.[19]

## *À la recherche du temps perdu*

On Friday, November 22, 1963, came the news that President John F. Kennedy was shot in Dallas. As the originator of media theory, Marshall McLuhan, said, in the worldwide electronic networks, we are now a "global village."[20] When an "event" of political significance or voyeuristic fascination happens somewhere, everyone everywhere knows about it in minutes. It could be a terrorist or natural catastrophe – or a Hollywood or golf celebrity arrested for drunk driving. JFK was murdered at 12:30 PM in Dealey Plaza. At 12:40 PM, the entire world knew about the shooting after Walter Cronkite of CBS News announced it on TV. Twenty minutes later, Kennedy was pronounced dead. Two days later, on live television from Dallas Police Headquarters, Jack Ruby shot and killed Lee Harvey Oswald, JFK's alleged assassin.

Monday was declared a day of national mourning. There was no school. My brother and I watched the funeral procession from St. Matthew's Cathedral in Washington, D.C. to the burial plot at Arlington National Cemetery. The haunting image of the shocked and grieving Kennedy family – his widow, his younger brother, the Attorney General, his younger brother, the U.S. senator, and his two children slightly younger than me – was inscribed into my mind.

CHAPTER 2

Members of the Kennedy family leaving the
U.S. Capitol building after viewing JFK lying in state
(Photo by Abbie Rowe, Wikipedia Public Domain)

The next day was Tuesday, and I was back at the Sands Point School. Classes were canceled, and we had free time to meditate on the tragic event and its implications for America. It was the shattering of an optimistic dream. It was a sign that tumultuous times were ahead – for me and my country. "Ask what you can do for your country," he said.

Mrs. Wagner, my caring teacher from the previous year, comforted me in my disillusioned sadness. We sat on a wooden ledge surrounded by the wet plummeted foliage of late Autumn, fallen from a now leafless tulip tree, overlooking the wild playground with its swings, sandboxes, tetherball poles, and tarpaulin-covered swimming pool.

> "You were right, Alan," she said. "The world is full of strife. And now you know what death is. But you still have your life to live. Make something of it. With the time that you have."

President Lyndon B. Johnson came to be well regarded in my child's mind for the signing of the 1964 Civil Rights Act, which, on some important levels, ended discrimination against Black people in America. Yet Johnson later became the arch-villain of the Vietnam War.

My final two years at Sands Point were relatively tranquil. Things were more relaxed for me in the fifth and sixth grades. I was past the scene of the crime. A temporary calm preceded the storm waiting for me when I re-entered public school in the seventh grade.

My fifth-grade teacher, Miss Carter, was as agreeable and caring as Mrs. Wagner. She taught us English and Social Studies. The fifth-grade class met in the psychically unspoiled territory of the mansion's third floor. Most of the kids in the class had discipline problems. They ran around all day and did not want to learn. Miss Carter found the time to give me individual instruction in English grammar. Her explanations and drills about the parts of speech and sentence structure gave me a firm foundation for writing and for learning foreign languages.

> "An adverb can modify an adjective," she said. "But an adjective cannot modify another adjective. And remember that the oft-used phrase "very unique" is incorrect. Something is either unique or it isn't. 'Unique' cannot be modified."

The gracious Mrs. Bernstein taught the same subjects in my sixth-grade class. Our class did not meet in the mansion but in the newly completed single-story annex building closer to the start of Elm Court. No bullies picked on me or gave me a hard time. I met other pupils who also skipped. It was not unusual to be younger than the standard age.

There was a girl named Patricia who had even skipped three years. She boasted that she surpassed me. "I skipped three times, so your twice is nothing!" she declared.

By fifth and sixth grades, I already intuited that it was all a big mistake. I also had a strong sense that there was nothing that I could do about it. My life was a done deal.

There were twenty pupils in the sixth-grade class. But, as chance and the finite number of American Jewish names would have it, there was another boy with precisely the same first and last names as me. Since there was no family name difference to tell us apart, the other kids called him "Big Alan" and me "Little Alan." That felt terrible.

My torturer for several months was a late teenager who traveled on the same minibus as me twice a day. Stanley was in the eleventh grade and was over six feet tall (183 centimeters). I was nine years old and four feet six (140 centimeters). For his sadistic pleasure, he stuck sharp needles into my legs every day. The van-sized bus seated fourteen passengers in four rows. The first-row seats were almost always taken. He always sat one seat in front of me, no matter which seat I sat in. He turned around with a big grin on his face and began his Nazi-like procedure. He did this to me and to another boy named Jon, who was also small. He pressed the pointed metal pins into my flesh above the knee and in various places on the front of my thigh. It was painful. On a forty-five-minute ride, he would do his stinging fifty times. I had to pretend that it did not bother me. Sometimes, I asked him in various ways to stop. But if you beg, that inspires the bully to intensify the torture. I asked the bus driver to intervene, but he did not want to. Talking to my parents was out of the question. It was better to endure physical and emotional torment than to go crying to Mommy.

If I had been with kids my age, I would have been good at team sports. Playing baseball, I was a good pitcher, fielder, and contact hitter. I could pass and catch an (American) football well. I had a smooth, on-the-money outside shot at basketball. I had a natural touch. But being two years younger and short meant I was always picked last when teams were chosen. I was relegated to being out of the action (for example, assigned to play right field and bat ninth in baseball). There were rare moments when I experienced my innate athletic ability in contests involving the participation of multiple individuals. In sixth grade, we played a lot of soccer. I was fast, energetic, and agile and scored many goals. The captain of our team was the best athlete in the upper elementary division. He liked me and sent me the ball often. It was the "once in a blue moon" when I felt not estranged.

Egerton Swartwout, the Beaux-Arts architect, built a six-car stucco garage or horse shed for the super-rich Luckenbach clan in the 1920s. It had six rooms, a bathroom, and living quarters on the second floor, presumably for the chauffeur or horse trainer and their respective families. This *dépendance* of the main house became the school's art classroom. My mother was a realist and slightly impressionist painter of natural and nautical scenes who paradoxically studied with the abstract master Mark Rothko, so I was encouraged to develop my artistic talent. The walking and car path to the art room house was a downward-sloping branch from the circular driveway. I made several flower vases from clay by pinching, making coils, and flicking paint colors with a brush at the solidifying object as it spun around.

At both ends of the vast schoolground, at the edges of the far eastern and western edges, were trails that went on for several hundred meters and led to the beach. The east path was less spectacular. You walked on an abandoned road of dirt and pebbles where some lespedeza and grassy weeds sprouted up. The vehicle passage, which had devolved into a footpath, and the Long Island Sound were on the same land level. The trail ended before you saw the beach. You had to fight through thorny bushes to get to the opening.

The pathway to the beach at the far west end of the property was the most enchanting site of my childhood. You passed through the gate of a metal fence that was often locked. You entered a wide arboretum strip with many species of

Florence Morrison (my mother), Fishing Village, 1963 (Alan N. Shapiro private collection)

trees and sculpted greenery vines overhead. It was filled with the sounds of birds and crickets as summer approached. There were plentiful honeysuckle flowers. This immersion in the sensorium of magnificent nature continued for many minutes as you neared the water. At the termination of the path, you arrived at high cliffs far above Long Island Sound. I ran down the steep slope of loose, moist sand in fearless joy, propelled forward at high speed by gravity, landing safely on the horizontal strand. When I later read J.D. Salinger's novel *The Catcher in the Rye*, it was that cliff I pictured in my mind's eye as the place where Holden Caulfield, in his fantasy, catches the children playing with abandon in the rye fields close to the edge where they, literally in the scenario, risk tumbling to injury or something worse, or, metaphorically, lose the innocence of childhood.[21]

Doing six years of elementary school in four years was the seed of the suffering. It set the stage for the challenge, enigma, and question mark of my life. Yet the question is not left unanswered. I write "to make lemonade" from the lemons.

The years at Sands Point were a paradise lost. I was free in utopia, and it gave me strength to face the misery that was to come. The lunch food at the school was good. I loved the roasted breaded chicken and the buttered flat noodles. My father attended the commencement ceremony. They awarded me the best math student in the graduating class.

CHAPTER 2

My father and I: 1966 Sands Point Country Day School sixth grade graduation
(Alan N. Shapiro private collection)

A few things about skipping twice truly interest me: Why was I never resolute enough to tell my parents that I was miserably unhappy and ask to be returned to the grade of children my age? What does the decision they made to mess up my life this way say about the political and cultural climate of the Space Age and Cold War 1960s? How was it possible for my parents to be so ignorant of psychology and what is today called emotional and social intelligence? What are the implications of narrating how one can be alienated in the capitalist and patriarchal society without being a "member" of a recognized oppressed group according to the usual categories of economic class, gender, race, sexual orientation, religion, and disability?

CHAPTER 3

# The Radical Left Late 1960s

A Hollywood movie that significantly impacted my psyche was *The Great Escape*, starring Steve McQueen, James Coburn, James Garner, Charles Bronson, and the rest of an all-star cast. I saw the film at the local Herricks Cinema movie theatre when it first came out in 1963. I was seven years old. I have re-watched it at least twenty times since then. The film gripped me emotionally and existentially because my father fought in World War II against the Nazis. It was unambivalently a "just war." The war was the immediate historical background to the suburban consumer-and-media-culture lifestyle that America's victory made possible. We won, and we were supposed to enjoy the spoils immensely.

On another level, the theme of escape – the desperate yet heroic effort by the baseball-loving McQueen character Captain Hilts fleeing from the Nazis on a stolen motorcycle and trying to daredevil-jump barbed wire fences into Switzerland – resonated deep in my mind.

"Los, aufstehen! Hände hoch!" the German soldier shouts at Hilts while pointing a rifle at him after the American flyer's motorcycle crashes.

To compare 1960s Long Island to Nazi Germany makes no sense whatsoever and would be a profound mistake. Yet, on an individual unconscious psychological level, the metaphor was reverberant. I desperately wanted to escape my situation of being two years younger than the students in my grade. I desperately wanted to escape from my mother, from her narcissistic personality disorder and her authoritarianism.

## Junior High School

I exit the front door of our house and turn right. I walk to the end of Gordon Drive onto the sidewalk of Herricks Road. I turn right and walk fifty steps – past the

CHAPTER 3

Steve McQueen in The Great Escape, John Sturges director, United Artists, 1963 (Alamy stock photo agency paid licensed image)

farm stand – to the corner of Center Street (a street with the same cookie-cutter houses as on our block).

I traverse the wide thoroughfare to the school side with the help of a female crossing guard who stops the traffic. I pass through a narrow parking lot where school buses unload their passengers. I go up seven steps to the main entrance to Herricks Junior High School, where I will be a pupil for the next three years.

There are three doors to the lobby next to each other, flanked by five tall decorative white Gothic pillars that are double the vertical extension of the doors. The school is a gigantic U-shaped building featuring innumerable brick walls all around. An American flag on a tall pole stands out front. Behind the school, there is a football field and tennis courts. There is a large gymnasium with wrestling rooms and a space suitable for basketball games.

In August 1966, a few weeks before my dreaded reentry into the public school system, my mother took me to a meeting with a guidance counselor. He rubber-stamped my enrollment without giving it a moment's thought. He nodded to my mother's request that I be put into the accelerated math class where I would already learn eighth-grade mathematics.

What followed were three years of endless humiliations. The others were twelve to fifteen years old. The boys entered puberty. Many had adult-like bodies. They acquired facial hair. Girls developed breasts. I was ten years old when I started seventh grade. I was, and looked like, a child.

I had no friends. No one wanted to be my friend.

No girl would speak to me or look at me. Other boys were popular. I was not. When the others approached me, they always said the same thing:

"What are you doing in our school? Why are you here?"

I had no answer. My strategy became to become as anonymous or invisible as possible. Let them look at me and see past me, see nothing. It was crucial that my age not be brought to anyone's attention. It should not become a topic of conversation. It would suffice that they think of me as small and young-looking.

I lived a daily battle for psychological survival. Every day, I dreaded going to school. There was no one with whom I could talk about my predicament, nor did I want to. My refuge was being alone after school, let out at 3 PM. Plus, the weekends, holidays, and the summer.

In my loneliness, isolation, and solitude, I walked around the neighborhood of fresh air, plastic hyperreal media, and consumer dreams. I looked longingly at everything non-human: the tall grass, the fire hydrants, the elevated water tower, the shelf items in the drugstore, and the crumpled trash in the gutter.

No one ever beat me up physically.

Every morning began with facing Robert, who sat before me in the homeroom class. He always gave me a hard time, made fun of me, bullied me, and insulted me.

"Hey, runt! You, half-pint! Pip-squeak!"

Robert was in my homeroom class all three years. The alphabet determined who was assigned to each of the dozen homeroom classes and in which chair and desk they sat. His last name began with the same two letters as mine.

Lunch was in the school cafeteria and cost 40 cents. The food was awful, but they had a good chocolate cake with frosting and cream filling for an extra 15 cents. The large, industrial-looking dining hall had many laminated round green tables. Having a group of friends with whom you could sit was essential. For a long time, I sat alone in embarrassment. Finally, some boys invited me to sit with them. At the end of the lunch session, you had to "bus" your tray, which meant bringing it to the food-tray-return conveyor belt, from where it would be sent to have its contents sorted by some kitchen employee in the back. There was a teacher observing everyone. You would be reprimanded if you left your tray on your table.

I carry my used tray to the station and return to my table. The other boys placed another tray in front of my chair, marking it my responsibility to "bus" that tray.

"You forgot to take up your tray!" they sneer.

CHAPTER 3

This cycle repeats itself multiple times. The boys laugh at me until the bell signaling the end of lunchtime rings.

One time, during an extended physical education period, the school administration brought all the kids from the eighth grade into the gymnasium. We were going to dance in pairs. All one hundred fifty boys lined up in order of height. There was a separate line of girls also in height order. I was the second shortest boy in the school. I was paired for dancing with the second shortest girl. We were the same height. The music started. We went through the dance motions quickly, then immediately and awkwardly went our separate ways.

In the locker room after gym class, I dreaded going to the shower. All the other boys had pubic hair already. I was eleven years old and did not. We played football or basketball. We participated in gymnastics or wrestling. At the end of the session, you were sweaty, and it was mandatory to take a shower. If these teenagers saw that I did not have pubic hair, they would taunt and mock me – and there would be questions about my age. It was twenty meters to walk from my locker to the shower room.

I take a towel and cover my crotch until the last possible second. I drape my towel over the metal hook in the side room. I go quickly into the shower. Several other boys are there, chatting with each other. I complete a pseudo-shower in a few seconds, exit, grab my towel, and cover my crotch again. I walk super-fast back to my locker.

Despite my misery and alienation, I was a good student. I got very good grades. My typical report card was four A's (English, social studies, math, and science) and one C. My C was always in French. I was terrible at French. Many years later, after living a year in France, I learned to read and speak French fluently. Eight years of school French taught me eight words.

I am in the eighth-grade French class. It is an unusually large class, about thirty-five pupils. It is early in the morning. Mr. Talbot, who is German, is the teacher. I retained nothing from yesterday's homework assignment. I sit in the back, praying he won't call on me.

Just as I have my glasses off and am rubbing my eyes to stay awake, Mr. Talbot calls on me. I don't even know what his French question was.

"That's Alan," exclaims Mr. Talbot wryly to the class. "He closes his eyes. He hopes that the world will go away!"

The male seventh-grade English teacher had us read seafaring adventures like Edgar Allan Poe's *The Narrative of Arthur Gordon Pym of Nantucket*, Robert Louis Stevenson's *Treasure Island*, Daniel Defoe's *Robinson Crusoe*, and Johann David Wyss' *The Swiss Family Robinson*.[22] When I said, "It is just like the TV show *Lost in Space*," he had no idea what I was talking about. He had no interest in, or knowledge of, the media of television.

# THE RADICAL LEFT LATE 1960S

Treasure Island Hotel Casino Las Vegas (Photo by Alan N. Shapiro)

Many years later, I played blackjack and roulette at the *Treasure Island* Hotel Casino in Las Vegas. *Treasure Island* is a story about pirate treasure. A secret and mysterious treasure is buried on a remote island surrounded by ocean waters. The immense treasure was not accumulated in the most upstanding and scrupulous ways. It was accrued by pirates who robbed and pillaged on the high seas.

The old brown seaman who stayed for a long while without paying at the Admiral Benbow Inn in the seaside village of Black Hill Cove tells of the treasure at Treasure Island. He arrived with his weathered sea chest containing all his worldly belongings, dragging it behind him on wheels. He was a rugged, undeniably masculine man.

> Fifteen men on the dead man's chest –
> Yo-ho-ho, and a bottle of rum!
> Drink and the devil had done for the rest –
> Yo-ho-ho, and a bottle of rum!

The old seaman, who called himself Captain Billy Bones, was perpetually on the lookout for a fellow seafaring man with one leg. This man was the fantastic character Long John Silver. Bones dreaded Silver's arrival. The young Jim Hawkins (the narrator and protagonist of the story) had nightmares where he imagined the notorious legendary creature in various horrific forms of the severed limb having been cut off at different levels of the lower body.

A strange sailor named Black Dog shows up at the Admiral Benbow Inn. At breakfast, the two former shipmates engage in a conversation that quickly turns hostile. A saber duel erupts. Black Dog is wounded in the left shoulder. Billy Bones collapses from a stroke, falling to the ground. He takes a fall. Jim and the Doctor carry Billy upstairs and lay him out on his sick bed. While Jim cares for Billy over a period of time, the old man confides in the boy that he once sailed the seas as First Mate to the pirate Captain Flint. The appearance of Black Dog suggests that Billy's former crewmates are nearby. They are here to confiscate his old sea chest.

Jim Hawkins and his mother hear the sounds of a group of rascals approaching. Seven or eight pirates, led by the blind man, Pew, descend upon the Inn in search of the chest. The pirates are not interested in the money but rather in the object Jim had removed from the chest: the oilskin bundle, which they call "Flint's fist."

Jim's adventure begins. He was going to the sea in a schooner, with a piping boatswain and pig-tailed singing seamen, bound for an unknown island to seek buried treasures!

The female eighth-grade English teacher thought the high-water mark of all Western literature was the Beatles' recently released album *Sgt. Pepper's Lonely*

Herricks Junior High School Math Award, 1969 (Photo by Alan N. Shapiro)

*Hearts Club Band*. She handed out copies of the lyrics of all the songs. We had heated discussions about who, or what, the "Lucy in the Sky with Diamonds" (1967) was that inspired that psychedelic rock masterpiece.

We had a typing class in eighth grade. I learned to type fast and accurately with my hands poised over the eight "home keys" – and my right thumb dangling over the space bar. It was an invaluable skill. Later in life, in my twenties, while surviving as a "starving artist" and "down and out" (in George Orwell's sense) in New York City, and not wanting to take a full-time job, I was able to get "temp" office work. I had to pass the "70 words per minute with five or fewer errors" administered by the Manhattan temporary work agencies.

Carrying out the instructions my parents programmed for me, I joined the school's math team. About once a month after school, we competed against the math teams of other North Shore of Nassau County middle schools. You had five minutes to figure out the answer to a challenging stumper. By the end of ninth grade, I was the highest-scoring competitor on the team. They awarded me the best math student in the graduating class.

## The Dreaded *Bar Mitzvah*

A narrow majority of the kids in the school district were Jewish. The major social events of the spring of seventh grade and the autumn of eighth grade were the *bar mitzvahs* (of the boys) and *bat mitzvahs* (of the girls). According to our religion, this passage happens when you are thirteen and "become a man" (or a woman). From that moment on, you are responsible for your actions. It was an occasion to get a lot of cash presents, usually in the form of U.S. savings bonds. Even more momentously, you could have a big party or "reception" at one of the local lavish country clubs. There would be amazing food, dancing, speeches, and reunions with all close and distant relatives. You could show off your popularity – how many friends you had, how many girls you knew, and who would accept your invitations. There was an informal competition regarding how many *bar mitzvahs* you were invited to.

The whole thing terrorized me. I had no friends, so I would not get any invitations. My *bar mitzvah* would not happen until the spring of ninth grade. It would be mortifying. There would be lots of attention drawn to the fact that I was two years younger. There would be no escape from it. Moreover, I had no friends to invite to my *bar mitzvah* party – and certainly no female friends. By the time the year 1967 ended, everyone else had had their *bar mitzvah*. Fortunately, no one noticed that I had not had mine. Mine was going to be in April 1969. That was a long way off. My strategy was to postpone anybody finding out about

CHAPTER 3

it as long as possible – and get through the event somehow. Once it was done, it would be forgotten.

Circumstances, however, threw a glitch into that plan. The date for my *bar mitzvah* was set. It was seven months in the future. I had to start to attend a *bar mitzvah* preparation class that met every Thursday at 5:30 PM in the school area of the synagogue. It took months to memorize in Hebrew the passages you were going to read out loud from the *haftarah*. You had to learn by rote prayers that you would lead the congregation at the Saturday morning ceremony. In that preparation class, I would be (in an exceptional circumstance) with a group of seventh graders: kids my age. In the daily life of junior high school, I was at the start of ninth grade.

For three years, I attended a late afternoon class in the temple once per week to learn Hebrew (I retained about twenty words from that). Once per week, on Sunday morning, I went to the temple for ethics instruction and Old Testament readings.

The *bar mitzvah* preparation class met in one of the four classrooms in the basement of the synagogue building. There was a big problem. Just before my class, there was a Hebrew- language class of ninth graders meeting in the same room from 4 PM to 5:30 PM. I knew four of the individuals in that class. They were not friends, but I had known them since the seventh grade. They were in the accelerated math class with me every year. They were in other classes of mine in junior high school. If those four adolescents saw that I was in the *bar mitzvah* preparation class, my being two years younger than everyone else would become a major topic of conversation. I had to avoid them becoming aware of it. Their Hebrew language class ended at 5:30. My *bar mitzvah* class began at 5:30 in the same room. I would be coming in just as they were going out. There was only one stairway leading down to the classrooms in the basement. There was seemingly no way to prevent them from seeing me.

But wait, there was a secret back stairwell! The large auditorium (not the main sanctuary) of Temple Emanuel was directly above the lower level with the four classrooms. One afternoon, the auditorium was empty. I went exploring. I went up on the stage. My intuition told me there must be an auxiliary stairway somewhere back there. There it was! To the left of the stage, behind a curtain, through a heavy door. From the landing, you could walk down two flights of fifteen steps each and come out just a few paces away from the fateful classroom!

The plan was clear: every Thursday after school, I would arrive early, walk through the auditorium, onto the stage, hoping that no one saw me, and enter the stealth stairwell. I would wait until 5:35 PM on the dot, then descend the stairs to the classroom. I would have the certainty that my ninth-grade schoolmates would have left, and my seventh-grade classmates would just be getting settled into the room and ready for the lesson from the cantor. It was the cantor who taught us how to sing Hebrew prayers.

For seven months, I spent twenty minutes every Thursday in my small, hidden cavity. The stairwell had no electric light, so it wasn't easy to see my wristwatch.

I sat on a cold, rough metal step. I was alone in the world. I trembled with fear. My heart was racing. My palms were secreting sweat. I belonged to neither group of pupils. I was betwixt and between. I heard voices from outside my backstairs cell. Children laughed, making rowdy disorder. I was hyper-conscious of the seconds passing. Tick-tock. Tick-tock. My wanting time to go faster made time go slower. Squinting to see my watch in the dim light seeping through an opaque window, I perceived that only one minute had passed since my last glance at the watch. What was I doing here? It was nasty, miserable, and lonely. Yet I understood something else about the circumstance replete with meaning. It was brilliant. It was genius. It worked. I found a solution – a way out of an impossible and desperate situation.

It was like Captain James T. Kirk of *Star Trek: The Original Series*, the only cadet at Starfleet Academy ever to solve the no-win scenario of the *Kobayashi Maru* fictional spacecraft training test simulation.

It was a sketch for future writing.

In the Sunday School class at Temple Emanuel that I attended while in ninth grade, there was a girl I was in love with named Gwen. During a portion of the two hours, we walked from the classroom into the main sanctuary for a brief religious service. We sat in rows of long pew-like lacquered wooden benches. I had the good fortune to sit next to Gwen several times. She was six inches (fifteen centimeters) taller than me. During the service, we alternated constantly between sitting and standing. Every time we stood up, I stood high on my toes, stretching my feet as far as possible for her not to notice how much shorter I was than her. It was very uncomfortable but a good idea.

One week before my *bar mitzvah* service, which was going to happen on a Saturday morning in April, it was my obligation to be present at the start of the *Shabbat* religious service on Friday evening in an honorary role as the boy about to reach his milestone of manhood one week later. One kid in my ninth-grade Earth Science class (at the junior high school) saw my name in the temple's weekly announcements.

In the middle of the class of twenty-five pupils, he rose and started to chant: "Alan Shapiro drinks the wine tonight! Alan Shapiro drinks the wine tonight!"

I didn't know what that meant. I hoped that no one else did either.

CHAPTER 3

Surprisingly, I was invited to three *bar mitzvahs* at the start of eighth grade. Two childhood playmates from Gordon Drive invited me to their ceremonies and country club receptions. Then there was Mark. We were in several classes together. He was from a family that was very liberal politically. Mark was empathetic to my plight. He took pity on me and invited me to his *bar mitzvah*.

I felt good in the *bar mitzvah* preparation class with children my age. I was *bar mitzvah-ed* on that Saturday in April 1969, paired with another boy born on the same day as me in 1956. We sat next to each other in that basement classroom. He was smart, kind, and empathetic. He immediately understood how unhappy I was in the ninth grade.

The dreaded day finally arrived. I sat at a table at my country club reception with four male companions: the two playmates from the street of my house, Mark the liberal intellectual, and another boy I befriended in the metal shop class. There were no girls. I was very embarrassed. The only thing that spared me from total humiliation was that Debbie, the twelve-year-old daughter of my father's business partner in his structural engineering firm, was there. I danced with Debbie. It felt good for me, and it felt good because people watched.

## Rebels Against the Establishment

In the late 1960s, the Vietnam War raged. It divided America and generations. Everyone had to decide if they were for or against the War. My brother and I were against it. My parents were for it. They felt it was their patriotic duty to support the War. In 1964, in the United States, the Free Speech Movement at the University of California at Berkeley began a rebellion that spread to university campuses across America, igniting a political movement against the War.

Paul Goodman became one of the intellectual heroes of the young generation. His 1960 book *Growing Up Absurd*, a critique of the state educational system expressed via his auto-socio-biographical experiences, made him a recognized writer of the New Left.[23]

The next Hollywood movie that massively impacted my psyche was *Little Big Man*, starring Dustin Hoffman. I saw that film at the Manhasset Cinema movie theater at the Miracle Mile shopping center on Northern Boulevard (next to the intersection of Port Washington Boulevard – the good old bus route to the Sands Point School) when it first came out in 1970. I was fourteen years old. I have watched it five times since then. I have read the book by Thomas Berger on which it was based.[24] The film deepened my opposition to the Vietnam War. It intensified my disenchantment with America. Via the story of Hoffman's character Jack Crabb, who was raised by members of the Cheyenne nation and who was present at "Custer's Last Stand" (the Battle of the Little Big Horn in 1876), the film depicts the genocide of the Native American civilizations by the white American settlers. I empathized with Hoffman-Crabb for being short of stature yet very brave.

Dialog in *Little Big Man* between Jack Crabb and General George Armstrong Custer:

> General Custer: "You don't look like a scout to me. A scout has a certain appearance. Kit Carson, for example. But you don't have it. You look like a mule skinner."
> Jack Crabb: "Well, I don't know anything about mules."
> Custer: "I can tell the occupation of a man merely by looking at him. Notice the bandy legs and strong arms. This man has spent years with mules. Isn't that correct?"
> Crabb: "Well I… Yes, Sir!"
> Custer: "Hire the mule skinner!" (Arthur Penn, director, Cinema Center Films)

At that point in my life, the soundtrack was the song *Eve of Destruction* (1965) by Barry McGuire:

> The Eastern world, it is exploding, violence flaring, bullets loading
> You're old enough to kill but not for voting
> You don't believe in war, but what's that gun you're toting
> And even the Jordan River has bodies floating
> But you tell me, over and over and over again, my friend
> How you don't believe, we're on the eve of destruction (Dunhill, RCA Victor)

On Sunday afternoons, my father often drove us to Manhattan in his car for activities like climbing bedrock in Central Park, visiting the Museum of Natural History, and eating dinner in Chinatown. We saw the movie *Robinson Crusoe on Mars*, the first science fiction film I loved, and *PT-109*, about JFK's heroics as commander of a torpedo boat in the Pacific theater during the Second World War.

My mother exhibited her paintings at the Washington Square Outdoor Art Show. She used her maiden name, Florence Morrison, as her artistic name.

On May 21, 1968 (I was twelve), I saw a dramatic multi-column headline in the *New York Times* that piqued my interest. I remember the exact place where I was when I saw it. I sat in the back seat of our car just as we exited the Midtown Tunnel – on our way back to Long Island – and just before passing through the toll booth. From the front passenger seat, my mother handed me the newspaper:

> FRANCE IS NEAR PARALYSIS AS MILLIONS JOIN STRIKE; REDS PRESS FOR COALITION. France headed toward virtual paralysis today as millions more of her workers occupied factories, mines, and offices.

It was the first time I heard about the student-worker near-revolution in May-June 1968 in France, which began at the suburban campus of the Université de Paris-X Nanterre and the Sorbonne. Students set up action committees and held assemblies. They built barricades from pavement stones, furniture, and cars turned on their sides. They fought against the police. The rebellion against the establishment was international. It happened in the foreign country with which my education made me the most familiar: France.

In my final year at the Sands Point School for Gifted Children, I went with a group of kids to a television studio in Manhattan to appear on Sonny Fox's *Wonderama* show on WNEW Channel 5. We met and had a group discussion with Senator Robert F. Kennedy, the younger brother of the slain President. I asked Bobby on camera if he thought the Supreme Court would need to take further action on civil rights for Black Americans now that the Civil Rights Act was signed into law. His answer tended towards a No. Bobby asked me if I wanted to be President when I grew up. I said No. He laughed and asked why. I replied that I did not want to "lead" anyone. He laughed again. "He doesn't want to lead anyone!"

After President Johnson's surprise announcement at the end of March 1968 that he would not seek reelection, I became an enthusiastic Kennedy for President supporter. Others, including my brother, believed Senator Eugene McCarthy was the more sincere and deserving antiwar candidate. I was attached to Bobby for emotional reasons. On the night of June 5, 1968, moments after winning the California presidential primary, RFK was assassinated at the Ambassador Hotel in Los Angeles. The shooter was Sirhan Sirhan, who was angry at Kennedy for his position on the Israeli-Palestinian conflict. The lackluster centrist Humphrey garnered the Democratic nomination. The mediocre Republican Nixon was elected to the Presidency. He continued the appalling Vietnam War for several more years, even expanding it to Laos and Cambodia. Once again, the loss of a guy named Kennedy from Massachusetts plunged me into profound disillusionment with America.

## My Love of *Star Trek*

I began in 1966, at age ten, to watch *Star Trek: The Original Series* episodes – together with my older brother Fred. *Star Trek* is a world where advanced science and technology have been deployed for the good of humanity. *Star Trek* is multiracial, multicultural, and multispecies. When NBC switched the show's time to 10 PM on Friday evenings for the third season, we held a marathon "lie-in" in Fred's room to pressure our parents to allow us to stay up until 11 PM. I lay awake thinking of the famous first-season episode "Shore Leave."

With the *Enterprise* crew needing leisure time following months of strenuous missions, landing parties, one of which is led by Dr. McCoy and Lt. Sulu, scout an uncharted Earth-like world to ascertain its degree of holiday suitability. The territory is plush and has a temperate climate. It has abundant trees, flowers, vegetation, and grassy meadows. Forests, rivers, and lakes are within perceptible range. The sensation of peace is "almost too good to be true." The weary Starfleet personnel who beam down enjoy a pastoral utopia. But this bliss is short-lived. It is disturbed by unexplainable happenings that intrude upon the waking dream of paradise.

McCoy tells Sulu that the place reminds him of Lewis Carroll's *Alice in Wonderland*.[54] A four-foot-tall white rabbit, wearing a waistcoat and carrying a gold pocket watch, appears. The rabbit announces that – my paws and whiskers – oh dear, I shall be late, yes, late for a very important date. Alice then appears. She runs into the clearing and asks McCoy if he has seen a giant rabbit with white gloves. The Doctor is dumbfounded but manages to point her in the right direction. She thanks him and follows the startling creature with pink eyes into the thicket and down into a hole in the ground.

Alice jumped into the rabbit hole without considering even for a second how she would get out again. She fell very deeply and for a long time, wondering if the well had an ending at all. She thought that she might be approaching the center of the world, or that she might even "fall right through the Earth," arriving at an antipodal country where people walk upside down.

In a fatigued mental state, with daydreams, stories, and free associations risen up to awareness, we are close to either a long drop into the depths of an alternate world with firm rules or poised for the poetry of illusions and appearances at the surface.

In my mind, I imagined my own "back story" to the *Star Trek* episode "Shore Leave" in the form of fan fiction:

> I was a first-class passenger in a mammoth spaceship, traveling from my planet of origin to the faraway Alpha Centauri A space colony.
>
> There, my first station as a grown-up awaited. I would take up my duties as a respectable functionary – in charge of some scientific discipline or other. Perhaps it would be a branch of medicine or engineering. There were limitless opportunities in those days for a young man of ambition, and I had years of spaceflight ahead of me for specialized training.
>
> *But then I saw a white rabbit.*
>
> It was a time of political upheaval and angry factional dissension among the citizens of our converted freighter. The journey took decades. Collective life aboard the immense vessel slipped into its own singular spacetime. Many light years separated the stars. We drifted.

CHAPTER 3

> I was part of a team assigned to scout a friendly-looking planet with no apparent life forms, gathering data with my tricorder to help determine its suitability for shore leave. I was alone momentarily, and there it was in the hedge. I had not swallowed any psychedelic red pill as in *The Matrix*, but the rabbit was the size of a man. He was sporting a necktie and carrying an umbrella.
>
> Perplexed about the cause of my sighting of the rabbit, I rubbed my eyes. He scurried off into the grove. I envisioned him moving through the underbrush of this mythical world. I did not follow him just yet. Dumbfounded, I motioned towards the cavernous dirt crater when Alice asked if I had seen a giant white rabbit come along.

After beaming down to the planet's surface, Captain Kirk lapses into a reverie about his younger days as a first-year Cadet at Starfleet Academy. While walking with McCoy, Kirk reminisces with bitterness about the "devil" who constantly picked on him – and who had Kirk so nervous that he never knew when or where his shadowing rogue antagonist would strike next. A grinning, broad-shouldered, twenty-year-old Irish lad, clad in an upperclassman Cadet's uniform, appears leaning against a tree trunk. The mischievous Finnegan has not aged since the two last met more than fifteen years earlier.

"You never know when I'm going to strike ya, ah Jim?"

Repeatedly calling the *Enterprise* Captain "Jimmy boy," the wily prankster baits Kirk into fisticuffs.

"Go ahead, lay one on me because that's what you always wanted. Isn't it? Come on, come on!"

*Star Trek: The Original Series*, episode "Shore Leave"
*(Paramount Global and CBS, Academic Fair Use)*

*Beckoning us suddenly from beyond the horizon of the system of functionality and rationality, memories come alive in an imaginary space and challenge us with unfinished business. They display the chaos and creativity of the artifice that can transform and alter us.*

> I lost my walkie-talkie link to the ship and instead heard the echoes of shouts in the trees, inflections I had not noticed, or words I could not make out. I saw my old nemesis, my personal Finnegan, racing through the glen.
> It was Stanley, the one who stuck the sharp needles in my thighs on the bus ride to school.
> Or maybe it was Robert, the one who bullied me in homeroom class.
> He was from school, and I owed him a day of reckoning. He leaped effortlessly over the jagged desert rocks. When I finally caught up and came face to face with him, it occurred to me I had no idea how he got here. Squinting my cheekbones, I felt strangely drawn to the unknown condition of his reappearance.

Now, my life's soundtrack changed to the song *White Rabbit* (1967) by Jefferson Airplane:

> One pill makes you larger, and one pill makes you small
> And the ones that mother gives you don't do anything at all
> Go ask Alice when she's ten feet tall
>
> And if you go chasing rabbits, and you know you're going to fall
> Tell 'em a hookah-smoking caterpillar has given you the call
> Call Alice, when she was just small (RCA Victor)

## John Glenn's Orbital Spaceflight

On February 20, 1962, John Glenn orbited the Earth three times. I was five years old.

Marine Lt. Colonel John Glenn, Jr. is aboard the capsule *Friendship 7*, a name symbolic of the camaraderie among the seven Mercury astronauts chosen to lead America's journey into outer space during the optimistic era of the Kennedy presidency in the early 1960s. Alan Shepard and Gus Grissom already went into space in 1961. Their capsules launched atop the much smaller Mercury-Redstone rocket. Their suborbital flights only lasted less than sixteen minutes. Glenn's orbital flight, known as the Mercury-Atlas 6 mission, is initiated atop the Atlas LV-3B, also known as the Mercury-Atlas Launch Vehicle.

Shortly before the start of the flight, there were repeated brief delays in the countdown. There were minor glitches with the microphone in Glenn's helmet and a bolt in the hatch door. After the gantry framework, which holds the rocket vertically erect, is retracted, Glenn sees a breathtaking view of Cape Canaveral through the periscope. Minus the scaffolding support, he feels the entire Atlas missile swaying side to side. As the liquid oxygen tank is filled, he feels the storage vessel and his body shudder together.

10 – 9 – 8 – 7 – 6 – 5 – 4 – 3 – 2 – 1 – 0. Glenn feels the engines start. 9:47:39 AM Eastern Standard Time. Liftoff at T plus 4 seconds. The spacecraft shakes noticeably. He feels a surge of velocity and knows he is in the air. After liftoff, the rocket rolls slightly to a right angle with the flight path, then pitches over to fly along the desired bearing. Glenn sees the Earth through the small mirror outside his window. Glancing upward, he experiences the horizon as a strange optical effect. He feels vibrations from the engines.

The two outboard booster engines are shut down and detached from the skyrocketing vessel. The noise decreases suddenly. Glenn sees a cloud of smoke but no fire. The escape tower is jettisoned. The sustainer engines are turned off. Acceleration decreases to zero. Glenn feels that he is tumbling forward. The clamp-ring between the Atlas rocket and the Mercury capsule fires. A loud noise. The force of the posigrade rockets separating the ship from the launcher gives him a raw feeling in his gut. The spacecraft turns around. Glenn sees the departed Atlas. It becomes a whitish object against the dark background of outer space, then a reddish object against the blue background of the Atlantic Ocean. He loses sight of the Launch Vehicle.

Glenn enters orbit at zero gravity. He contacts the human NASA Capsule Communicator at the Bermuda range station. He receives instructions on when to fire the retro rockets. He checks all control systems. During the first orbit, the spacecraft is on autopilot. Following a malfunction of a low-torque thruster, Glenn switches to manually guiding the capsule. He must sacrifice time planned for scientific experiments to the priority of survival.

The 1983 Hollywood movie *The Right Stuff* depicts the stories of legendary U.S. Air Force test pilot Chuck Yeager, the Mercury 7 astronauts, and John Glenn's orbital flight. The film is based on the same-named book by Tom Wolfe. According to the cinematic version, Glenn (played by Ed Harris) was spiritually accompanied during much of the mission by fellow astronaut L. Gordon "Gordo" Cooper (played by Dennis Quaid). Cooper engages Glenn in conversation as Capsule Communicator via two-way radio from his range control station in Muchea, Western Australia.

As he travels over Australia, Glenn sees the tribute created by the cities of Perth and Rockingham. They turn on house lights and streetlights as the *Friendship* 7 capsule passes overhead. A group of Australian Aborigines lights a fire in the open air in honor of Glenn. The native inhabitants chant, play woodwind music, and send sparks of fire into the sky in a symbolic communion with Glenn. Outer space travel is cryptically related to the reassertion of suppressed non-Western cultures.

While traveling over the Pacific Ocean at the break of dawn, nearing the end of his first orbit, Glenn sees small luminescent particles outside the capsule, which he describes as "fireflies dancing around." Mission Control on the ground at Cape Canaveral sees the warning light LANDING BAG DEPLOYED flashing, indicating a possible malfunction of the heat shield having come loose. It might be that the landing bag has been stretched.

Glenn would burn up and die if the heat shield fell off the spacecraft before or during reentry. The decision is made to reenter the atmosphere with the retrorocket package on. Its strong straps can keep the possibly loose heat shield in place.

Glenn begins reentry after three orbits or five hours in space. As he falls to Earth with his capsule burning up, he hums *The Battle Hymn of the Republic*. He brings the spacecraft down with his manual piloting skills. During the "ionization blackout," radio communication with Mission Control is interrupted. Only when visual contact is reestablished will the support team know that either Glenn is safe or the capsule has disintegrated. Glenn sees a glowing orange color outside. Flaming fragmented pieces from the retropack have broken off and flown past the small window. It becomes intensely hot for Glenn inside the capsule.

Splashdown and Recovery, 14:43:02 military time, 800 miles southeast of Bermuda. Recovered by the destroyer ship *U.S.S. Noa*. In the next scene of *The Right Stuff*, Glenn is waving to the massive crowd from an open car during the celebratory ticker tape parade down Broadway in lower Manhattan, New York City.

John Glenn (Wikipedia Public Domain),
Yuri Gagarin (Alamy stock photo agency paid licensed image)

CHAPTER 3

## Yuri Gagarin's First Manned Spaceflight

In 1961 – a few days before my fifth birthday – Air Force Major Yuri Gagarin of the Union of Soviet Socialist Republics circled the planet. On April 12, backed by scientists, designers, engineers, technologists, doctors, and military personnel, Gagarin achieved the first human spaceflight in history in the Soviet spaceship *Vostok I* (launched from a modified ICBM rocket). He orbited the Earth one single time and made a safe landing on dry land.

Gagarin was a proletarian by background, the "salt of the Earth," a worker from a family of farmers. His father was a peasant who later became a skilled carpenter. His mother loved to read books. Gagarin received formal education as a military flyer. He was sent to the Luostari airbase near Norway. In 1957, he became a Lieutenant in the Soviet Air Force. In 1960, he was chosen, along with nineteen others, for the Soviet space program. He then became one of the original Vostok program cosmonauts known as the Sochi Six.

The first human being in outer space was not an American, not Alan Shepard nor John Glenn. It was a citizen of a nation that no longer exists, a country that has disappeared off the face of the Earth.

The *Vostok* space capsule carrying Gagarin had no cameras for video transmission or photographing images of the Earth. It was equipped with only two cameras sending back television images of the cosmonaut: one camera showing a frontal view of his face, the other a profile view. Western media culture's *spectacle of the image* was absent.

At the spaceport of the Baikonur Cosmodrome, Gagarin climbs the stairs into the elevator that will rise diagonally to take him up to his spherical capsule atop the rocket. Before entering the lift, he waves and smiles at his comrades and friends below. He goes inside the spaceship. One hour later, the signal for the start of the rocket's engines is given. The camera in the cabin is switched on. The fueling tower moves away from the side of the rocket. Ignition. The power cable swings away. Blastoff! The supporting arms retract. The rocket is released.

"Here we go!" shouts Gagarin. He feels intensely the acceleration and vibrations as they penetrate his body. Due to the G-force, he can no longer move his arms and legs.

A beautiful confusion of lights and colors. The clouds part as he pierces the sky. Gagarin passes through the dense layers of the atmosphere. He sees the Earth shrouded in haze. The *Vostok* achieves an altitude of more than 300 kilometers. Gagarin flies over South America. He sees the details of Earth's topography and landscapes. He sees the seas, mountain ranges, cities, large rivers, and tracts of forest. He sees the continental coastlines and islands. He can distinguish plowed fields from meadows. He can move his arms and legs again, but they have no weight. Objects float in the cabin. Gagarin rises from his seat and is suspended in the air.

Ground stations measure the parameters of motion of the spaceship, receiving telemetry data during the flight. Information is transmitted via communication links to computers in data processing centers. Before deceleration, the spaceship's orientation system, using a series of optical and gyroscopic sensing devices, turns the axis of the capsule in the direction of the sun. Over the west coast of Africa, the deceleration motor is fired. These operations are controlled by software. The ship settles into a re-entry trajectory. The landing system is activated. Ninety minutes after takeoff, Gagarin sees the Volga River and the city of Saratov in European Russia.

Gagarin leaps with his parachute, ejecting at 23,000 feet (7000 meters). He comes to rest in a pasture. A woman and a girl see the blackened steel capsule coming down. The capsule lands in a field. Gagarin greets the two civilians. Tractor drivers approach with their vehicles. A helicopter appears with members of the spacecraft landing support team. They take Gagarin to the nearest town. He speaks by telephone with Khrushchev.

Yuri Gagarin's dream was to fly to the Moon, Venus, and Mars. Instead, he was banned from spaceflight by the Soviet Space Program because the nation did not want to risk losing its treasured Hero of the Soviet Union in an accident. Gagarin was killed, along with flight instructor Vladimir Seryogin, on March 27, 1968, in a crash of a MiG-15UTI plane.

On July 21, 1969, Neil Armstrong stepped onto the moon's surface. I watched it live on television. I was at the summer camp that I attended in New Hampshire. I went to that camp for four summers, ages ten to thirteen. For eight weeks, all the campers watched no TV at all. But we gathered for this global media event in the recreation cabin to see the astonishing images. JFK's promise that we would reach the moon before the end of the 1960s was fulfilled.

## Summer Camp in New Hampshire

In my child's mind and emotions, I experienced summer camp as a caste system like the Hindus in India or the feudalism of medieval Europe, both of which we learned about in Social Studies class. Limiting myself to the last of my four summers in New Hampshire, I will now tell anecdotes from that summer when I was thirteen, between ninth and tenth grades.

The headquarters of the camp enterprise was in Roslyn on Long Island. The owners also had a summer day camp in Roslyn. Day camp was like going to school, except you played sports and games instead of learning educational subjects like during the other three seasons. You spent the night in your own home. I attended the day camp one summer – before getting "promoted" to the "sleepaway" camp. To get to the camp in the mosquito-friendly New Hampshire woods and lakeside setting, you took a short plane flight on Eastern Airlines from LaGuardia Airport in Queens to Boston's Logan Airport, followed by a long bus ride north.

There were boys' and girls' sides of the three-hundred-acre camp. On the boys' side were fourteen residence cabins, also known as "bunks," divided into five Groups of two to four cabins each. You did sports and activities with the campers in your Group. The campers ranged in age from seven to fifteen. The fifteen-year-olds were in Group Five. In my final summer, I was in Group Four in the pine-floored bunk called *Kancamagus* (named after the seventeenth-century chief or *Sagamore* of the Penacook Confederacy of Native American tribes of southern New Hampshire). The previous summer, I was in the bunk called Devil's Den.

There were twelve campers in each bunk and four bunks in Group Four. Forty-eight kids were my fellow male campers with whom I interacted. I convinced my parents and the camp's owners to place me with kids one year older than me rather than two years older, as was my situation in school. The age and size difference between me and the other kids was less pronounced than during the scholastic year.

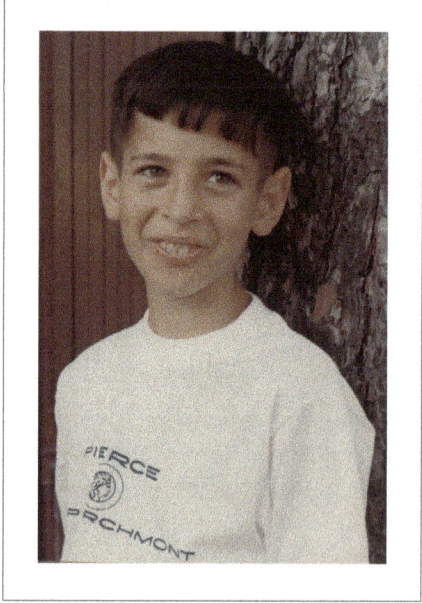

Pierce Camp Birchmont, East Wolfeboro, New Hampshire, age 13
(Alan N. Shapiro private collection)

It was not, strictly speaking, a Jewish camp, although *de facto*, almost 100 percent of the campers were Jewish. The owners were not Jewish. It was not, strictly speaking, a tennis camp, although tennis was the most important sport. Competitive tennis determined the "pecking order" of who was cool or not. It was the measure of your social standing, acceptance, or popularity. Tennis established your ranking in the caste system. Baseball/softball was the second most crucial sport for deciding coolness, followed by competitions like basketball, volleyball, and swimming. Most of the kids reveled in learning how to shoot a rifle, but I was not fond of that, drew a line in the sand, and refused to participate in that pantomime of violence.

You arrived at the camp at the beginning of July and stayed until the end of August. Your parents were happy to get a break from you for eight weeks. After four weeks, Mom and Dad visited for a weekend. Tasty barbeque food (hamburgers, hot dogs, chicken, fresh corn) was served to keep the parents from becoming directly aware of the blandness of the usual food.

Once during the summer, on a Sunday, we were served bagels, cream cheese, and unsalted Nova Scotia lox for breakfast as a special treat. As a prank, one of the counselors led a group of us at six AM to the camp kitchen to pilfer the bagels and lox. The camp owner got on the booming loudspeaker and warned the unknown thieves harshly.

We paraded around the sports field carrying the heavy food coolers and chanting: "We cannot tell a lie! We cannot tell a lie! We stole the bagels! We stole the bagels!"

Tennis at all levels worldwide is an ultra-competitive sport focusing on rankings and tournament seedings. But the system can be "rigged." Climbing the hierarchical ladder can be impeded by factors that work against fairness. All forty-eight campers in Group Four were semi-officially ranked in order of their alleged skill at tennis. I was a pretty good tennis player, although not great. My forehand was excellent. My backhand was weak. I often "ran around" my backhand. I could never figure out how to serve correctly, so I had to lob my serve. I would say objectively that I was about the eighth- or tenth-best player among the forty-eight. But it took me until the end of the summer to get recognition for that. One thing that held me back was that I could never persuade my parents to buy me a metal racket. The better players had metal rackets, but I was stuck with the inferior wooden variety.

The camp had three sets of outdoor tennis courts, with four hardtop courts in each set. The property was set on alternating slopes and plateaus, so the courts were labeled as the Upper, Middle, and Lower Courts. We had a tennis hour or two almost every day. At the beginning of the summer, all the boys in Group Four were assigned to play on one of the three sets of courts. The decree was for the duration of all eight weeks. Each of us was classified according to his supposed tennis skill as admirably excellent, acceptably average, or ineptly terrible.

On the second day of the summer, the head tennis counselor observed me for thirty seconds hitting three shots, then classified me as a miserable Lower-Court loser.

"This boy is a Lower Court tennis player!"

It was all about the simulacrum of reputation – the model precedes the real. And about lords and serfs. The "in" group of self-proclaimed better athletes in our residence cabin wielded power over the rest of us in the "out" group. That "clique" of arrogant boys called themselves the "Fab Five," deriving their moniker from the nickname of the Beatles. Instead of John, Paul, George, and Ringo, it was Jerry, Richie, Billy, Harry, and Charles Joseph "Chuck" Jr.

"But Alan, I thought the Beatles signify only the greatness of the 1960s!"

Chuck was the only non-Jew in our bunk. He was a cousin of the camp's owner.

Relegated to the Lower Court, I was only allowed to play against mediocre players whom I could easily beat, victories which could add nothing to my reputation. I asked a few times to be moved up to at least the Middle Court, but the request was denied. Another boy unhappy with his Loser label was granted his petition to move up, but he was an "old timer" who had attended the camp twice as many summers as I had. His family was connected to the owners. He performed a lengthy crying session, which pressured the tennis authorities to yield.

The same tennis counselor who had sealed my fate later observed me further and concluded that I had some tennis talent. We competed against another southern New Hampshire camp, and he decided to deploy me in his devious cheating scheme. He put me on the team and pretended to the opposition that I was one of the best players in our camp. He "stacked" me as the second singles player in the competition. I would lose, of course, but that was a pre-calculated cost built into the ruse. The trick set things up for him to play our second-best player against their third-best player, our third-best against their fourth-best player, and our fourth-best against their fifth-best player. We had an edge in every match except mine.

Everyone expected me to lose.

"Point game set match!"

When I surprisingly won, my so-called cohorts pretended that it had never happened. Some on my team accused me of cheating.

I was never officially reassigned from the Lower Court. But as the summer went on, I was allowed to play against better players. They entered me in the end-of-summer "A" tournament. I won two matches and made it to the quarter-finals, where I lost a close contest to the number three seed.

There was a similar social hierarchy or food chain on the girls' side of the camp, although I did not understand it. I guess at the top of the hierarchy were the girls who were both excellent athletes and physically attractive. If you had only one of those two qualities, your socially and existentially estimated worth was probably somewhere in the middle.

During the third week of the summer of 1969, there was a "social" dance in the recreation lodge cabin building. The Group Four boys and Group Four girls came together that evening. I stood shyly with my back to a wall, listening to the ubiquitous pop-rock music. A pretty girl named Laurie came to me and asked me to dance. Like me, she was not tall. We danced for maybe a half hour. Then we chatted for a while.

"I'm not a good athlete," she told me. "I like you a lot, and I want to be your girlfriend for the summer. You're more sensitive and smarter than the other boys."

The next evening, the recreation hall was set up for a movie, and we could meet again, sit next to each other, and hold hands.

My life's soundtrack changed to the song *I Think We're Alone Now* (1967) by Tommy James and the Shondells.

> Running just as fast as we can, holding onto one another's hand
> Trying to get away into the night, and then you put your arms around me
> And we tumble to the ground, and then you say
> I think we're alone now; there doesn't seem to be anyone around
> I think we're alone now; the beating of our hearts is the only sound (Roulette label)

Back at the Kancamagus residence cabin the following morning, everyone looked at me scornfully. They all saw what happened. A non-Fab Five member was not supposed to get a pretty girl. It was not going to be allowed. The only person to congratulate me was a counselor named Steve Aronowitz, who was about twenty and a Detroit Tigers fan.

When I arrived the second evening at the recreation center for the movie screening, Laurie was nowhere to be found. While I looked around for her, the place filled up, and there were no longer two free seats together. Her cabin mates had clearly conveyed to her that the budding romance with me must be aborted. When I finally found her sitting with a group (she was hiding from me), she acted happy to go with me, but it was too late to find seats.

I politely asked a few individual boys if they would mind moving over one seat to make room for us, but they all refused with disdain. The lights went out, and the film started. Laurie and I went our separate ways. I went to sit in the

Lonely Losers section on the floor in the front, near the screen. I never spoke with Laurie again.

A few weeks later, at another dance social in the same lodge building, I saw her sitting next to a boy from another bunk of Group Four who was widely admired for his athletic prowess. They sat in silence and appeared to have nothing to say to each other.

I have often thought of this story with sadness and regret. However, the wise lesson learned from the experience is the opposite. Alan, you were fortunate that it ended after that one evening! What would have happened if you had found two seats together watching that movie? It would have been complications and misery.

Against all odds, you achieved the unforgettable one evening of dancing with her, like Yuri Gagarin's one orbit around the Earth. The thirty minutes with her were more valuable than a long, hot summer. It's not a memory. It's something eternally living.

Perpetual radio music accompanied my summers in the late 1960s. Hit tunes recorded by the media of the burgeoning popular song industry played a massive role in my consciousness. There were songs of love, of political protest, and of philosophy:

*Happy Together* (1967), by The Turtles

Imagine how the world could be, so very fine, so happy together
I can't see me loving nobody but you for all my life
When you're with me, baby, the skies will be blue for all my life

Me and you, and you and me, no matter how they toss the dice, it had to be
The only one for me is you, and you for me, so happy together (White Whale label)

*In the Year 2525* (1969), by Zager and Evans

In the year 2525, if man is still alive, if woman can survive, they may find
In the year 3535, ain't gonna need to tell the truth, tell no lies
Everything you think, do, and say is in the pill you took today

In the year 4545, you ain't gonna need your teeth, won't need your eyes
You won't find a thing to chew, nobody's gonna look at you (RCA Victor)

## The Jets, Mets, and Knicks

On January 12, 1969, the team I root for in American football, the New York Jets, achieved perhaps the greatest upset in American sports history. The Jets were the champions of the American Football League (AFL). They defeated the powerhouse Baltimore Colts, the National Football League (NFL) champions, in

Super Bowl III. The Jets were underdogs by a wide margin, given no chance of winning. They were led by the brash rebel Joe Namath, who, before the game, guaranteed victory. The contest was also historic in that it legitimized the merger between the two leagues (the AFL and the NFL).

The Jets' victory made it seem like all great things were possible and that underdogs could triumph. But the win was a pact with the devil. Since then, the Jets have been a historically terrible team. They have never made it to the Super Bowl again.

In the 1960s, the team I root for in baseball, the New York Mets, was one of the worst teams in baseball history. They were an expansion team in 1962 and finished in last or next-to-last place for their first seven years. They never had a winning record. But in 1969, they suddenly had a great team. They were underdogs by a wide margin in the World Series against the powerhouse Baltimore Orioles, given no chance of winning. Led by the pitching of lefthander Jerry Koosman and incredible catches in the outfield by Tommie Agee and Ron Swoboda, the "amazing Mets" were triumphant.

The Mets' victory made it seem like all great things were possible and that underdogs could triumph. But the win was a pact with the devil. Since then, the Mets have been a historically terrible team. They have only won one more championship.

In the 1969-1970 season of the National Basketball Association, the team I root for in basketball, the New York Knicks, won their first championship. In the finals of the playoffs, they defeated the powerhouse Los Angeles Lakers. Led by their captain Willis Reed, talented point guard Walt "Clyde" Frazier, and Bill Bradley, who would later become a U.S. senator and a presidential candidate, the Knicks were the third New York "team of destiny" in the same year. In the decisive seventh game of the finals, the injured Reed came out on the court and played despite intense physical pain. He willed the team to an inspiring triumph.

But the win was a pact with the devil. Since then, the Knicks have been a historically terrible team. They have only won one more championship.

CHAPTER 4

# High School

April 22, 1970, the day before my fourteenth birthday, was the first Earth Day. I participated in my first political activity, attending environmental workshops and teach-ins at Herricks High School about air and water pollution, oil spills, and endangered species. We took long walks, planted trees, and did "cleanups" of the expansive grass fields. Inspired by Rachel Carson's 1962 book *Silent Spring*, environmental reform became a political issue.[26] Consciousness about the potential danger of Earth's ecological devastation and protection of the natural ecology of humanity's planetary habitat was beginning. Attention was starting to be paid to global warming, climate change, the destruction of the ozone layer, the melting of the polar ice caps, deforestation, excessive oil drilling, and loss of biodiversity.

During most of the 1960s, my father had a middle-class income. Around 1967, following the death of his boss, he became a 50 percent partner in his consulting engineering firm. His income took a sudden jump upward. In 1970, my mother decided that we should move to a much fancier house. The new upscale house was in the upscale Country Estates housing development of the Village of East Hills in Greater Roslyn on Long Island. The new house was only eight kilometers (five miles) from the old house, but I had to switch schools. Hence, I only attended Herricks High School for one year, in the tenth grade. I did my junior and senior high school years (eleventh and twelfth grades) at Roslyn High School. The families of the Roslyn kids were more affluent than in Herricks. There were also African American kids in Roslyn. There were none in Herricks.

To get to Herricks High School, I walk in the same direction as the Junior High School. I go on for ten minutes through a sports field and up a hill.

In tenth grade, at ages thirteen and fourteen, I had the first genuine friend of my life. His name was Jerry. He was the first person who ever wanted to be my friend – who saw valuable qualities in me, looking past the superficial fact that I was only five feet (152 centimeters) tall. Jerry was nearly six feet tall and three years older than me, but neither our height nor our age difference mattered to him. We talked a lot about life. I watched some of the 1969 Mets World Series games at his house. Some of Jerry's other friends wanted nothing to do with me – since I was an outcast and essentially an "untouchable," to invoke the Hindu caste system term. Jerry was also a very sensitive and thoughtful outcast in his own way.

The friendship with Jerry sustained me emotionally and psychologically during that critical "bridge" year of tenth grade, so I am very grateful. Our friendship was fervent and profound. We did things together almost every day. Yet the duration of the friendship was only one year, like the one Earth orbit of Yuri Gagarin's spaceflight. In June, at the end of tenth grade, my family moved to a new house in Roslyn. I knew I was done with the original scene of my childhood and Herricks High School. It seemed expedient at that conjuncture to break off my short-lived super-intense friendship with Jerry. I ended the relationship because I foolishly internalized the cult-like ideology of "high intelligence" and accomplishment. Jerry was not quite at my test-measurable smarts and scholastic aptitude level. I thought associating with him as a friend would lower my status or damage my prospects for success in the competitive achievement system. It was a big mistake by me. I know that Jerry suffered from losing our friendship. A few times over the years, he contacted me and tried to persuade me to change my mind.

The French class in tenth grade was challenging for me daily, and I could barely keep up. I sat at the second desk from the front on the left side, next to the windows. Once again, a girl was sitting immediately in front of me with whom I was in love and who was "out of my league," as the saying goes. An interesting aspect of that tenth-grade French class was that the teacher had us read the classic book *Village in the Vaucluse* by Laurence Wylie, a fantastic description of everyday life in a small French town in Provence after the Second World War.[27] Vaucluse is the department of which Avignon is the capital. The book gave me a vivid picture of cultural life outside America and postmodern consumerism.

There was no accelerated math class for tenth graders, so they put me in a geometry and trigonometry class for eleventh graders. I was fourteen years old to their age seventeen. It was a large class of about twenty-five teenagers. I must admit that I enjoyed it. The fact that I was not in their grade or institutionally designated peer group made the situation lighter, even comical. It removed the usual resentment or perverse curiosity about who I was – and what I was doing in the grade I was in. I had by far and away the best grasp of trigonometry of anyone in the class. I was the master of sines, cosines, and tangents. I got 100, or nearly 100, on every test. I sat in the front surrounded by three lovely and amiable seventeen-year-old girls. I was a child to them, but they adopted me as their "mascot."

The tenth-grade English teacher was a brilliant left-liberal woman who made great commentaries on the novels and plays she assigned us to read: *The Grapes of Wrath* and *Of Mice and Men* by John Steinbeck, *Death of a Salesman* and *The Crucible* by Arthur Miller, and *The Great Gatsby*.[28] Steinbeck made me feel empathy for people suffering in poverty. Miller poignantly portrayed social and individual alienation and political injustice in dramatic form. Fitzgerald sensitized me to the moral corruption that often accompanies extreme wealth. I was affected by *Black Boy*, Richard Wright's memoir or auto-socio-biography of growing up in the South, and *Invisible Man* by Ralph Ellison.[29] My own subaltern experience made me empathize deeply with the existentialist struggles of the black victims of racism. Perhaps this is an absurd statement to make, given my white male privileged socio-economic status in American society.

## My Hometown is Roslyn, Long Island, New York, USA

I lived in Roslyn for two years, from age fourteen to sixteen. I left when I was sixteen. I returned many times in the summer and during vacations. Nonetheless, I feel that Roslyn is my hometown. I knew Roslyn from visits to the "famous among children" Roslyn Duck Pond in nursery school and kindergarten.

In 1980, I sat inside the Bryant Library in the "haute bourgeoise" town of Roslyn (next to the duck pond). I read all the books they had about the student-worker near-revolution in May-June 1968 in France: books like Alain Touraine's *The May Movement: Revolt and Reform* and Alfred Willener's *The Action-Image of Society*, both of which were translated into English and brought out by prestigious American publishers.[30] These insightful historical studies deepened my appreciation for the enlightened ideas of the "spirit of 1968." The Bryant Library was built in 1876 by the poet and newspaper editor William Cullen Bryant. It is the oldest library on Long Island.

In 1614, Dutch trading ships traveling from the region known today as the state of Connecticut landed on what is now called North Hempstead Beach Park (on Hempstead Bay) and first visited the location that would become Roslyn. There was a Native American settlement in the territory for many centuries. In 1644, English settlers under Dutch rule took up residence in the area and created the township of Hempstead. In the 1650s, the Matinecock Indians, a branch of the Algonquin Nation, were drastically reduced in population by illnesses. Colonists captured their last villages. In 1664, the English took control from the Dutch.

## CHAPTER 4

In 1790, President Washington visited Hempstead Harbor during his tour of Long Island. On September 7, 1844, the name of the Hempstead Harbor Post Office, which is behind the duck pond and the wetlands, was changed to the Roslyn Post Office. Hempstead Harbor became Roslyn.

Roslyn High School was founded in 1904. The structure at its more permanent location on Roslyn Road was built in 1925. The old wing was torn down in 1971 while I was a student.

Twelfth grade was a great time. After all that I suffered during most of my school years, it was the first great time of my life. Before I recount my surprisingly upbeat experiences of twelfth grade, I will tell some things about my relatively uneventful eleventh grade, my first year in Roslyn. The eleventh grade will flow into the final high school year in the narration.

Like the old house in Herricks, I walked to and from the school from our new house in Roslyn. It was a downhill stroll, then an arduous trek back up the hill. I was allowed to go home during the lunch period. I ate a sandwich and canned soup.

The eleventh-grade English and Social Studies classes were mediocre. They were better at Herricks High School. Nearly nothing happened. They had us read *Beyond the Melting Pot* by Nathan Glazer and Daniel P. Moynihan.[32] The book was presented as a great work of social science that argued that the "melting pot" metaphor did not match with the retention of ethnic identities of the Puerto Ricans, Jews, African Americans, Italians, and Irish of New York City. It seemed to me, however, that something like the melting pot metaphor operates well at another level. Italian Americans have become much more American and less like Italians in Italy. American Jews have developed their own very American version of Jewish identity. Something like the "melting together" into a shared American culture can be understood as a system of "simulated" differences within sameness, or sameness with differences, rather than being the claim to complete elimination of ethnic identities in a homogeneous monoculture and the concomitant easy refutation of, or multicultural opposition to, that thesis.

In eleventh grade, I finally got straight As. I erased my French failures by switching to a beginning Spanish class. I became a top Spanish student, starting from scratch with a new foreign language. I was also the best student in the science class, which was chemistry.

I fell in with a group of boys who played poker on Friday nights. I was not very good at poker and usually lost about ten or fifteen dollars. However, one time, everything went my way, and I won several large pots. The older brother of one of my usual cronies was playing with us for the first time. I felt bliss while winning one hand after another. I was finally going to get back some of the money I had been losing consistently for months. We played with chips and then exchanged the money from losers to winners at the end. When it came time to "settle up," it was calculated that I had won fifty dollars, my friend Rick's older brother had lost fifty, and everyone else broke even. It now fell upon the older brother to pay me the fifty dollars.

He rose calmly from the table and strode to a sofa in the TV room. He was eighteen and big and fat. I was fourteen and short for my age. "You know what?" he said with self-satisfaction, "I'm *not* going to pay you!" It was a gloomy turn of events. I hoped that one of my so-called "friends," perhaps the younger brother Rick, would take my side and make plain to the big guy that he had to pay me. But no help from any supposed ally was forthcoming. They regarded my predicament with cruel amusement. Short of grabbing a kitchen knife and stabbing Rick's brother in the chest, there was nothing that I could do. In hand-to-hand combat, I surely would have no chance against the much larger opponent. I did nothing. It was 11 PM. I phoned my mother and told her to come and pick me up in her car.

I became a member of the community of the smartest kids in the grade, mostly males, centered around the accelerated mathematics class. They accepted me ambivalently into their scene. The atmosphere in the group was highly competitive camaraderie. The sense of rivalry intensified the following year in the twelfth grade. There was competition for grades, achievement, and genuine comprehension in math and physics. There was also a spirit of questioning of dominant American values, inspired by rock music, prose and poetry literature, and political protests of the youth culture of the 1960s and 1970s. In subsequent years, I observed that, for my former high school friends and classmates, that attitude of cultural contestation had been short-lived.

There was competition over who had the highest grade-point average and would thus be nearest the top when the final class rankings were announced. There was competition over the highest SAT (nationwide Standard Admission Test in math and English) scores. In eleventh grade, we took the PSAT test, a rehearsal for the real thing. It gave you information about where you stood and where you needed to improve. My math test score was nearly perfect. My English score was not good. This resulted from losing my desire to read books voluntarily after being forced to read the impossible *The Last of the Mohicans* at age seven. I did special tutoring for vocabulary expansion (SAT Prep Course) on Saturdays. My English test score went way up.

Above all, there was jockeying for positions over which colleges you applied to and to which colleges you were accepted. Who among us was going to Harvard and the other Ivy League universities? Applications for "regular decision" were to be sent in by mid-January, and the college admissions offices would notify us of their verdicts on our candidacies in April.

In the extracurricular activities I participated in, there was vying for superiority over who was the best contract bridge player, the best chess player, the best at skiing, and the best table tennis player. I was very good at contract bridge but not good at chess. I also formed some friendships while engaging in these pursuits. I played in the table tennis club. I was very good at "ping-pong."

There was a round-robin table tennis tournament at the end of twelfth grade. I played well and notched a series of wins before losing in a runoff final. It was a great sense of triumph and achievement for me.

In my match against Steve, a friend and a competitive rival in many things, I am one point away from losing the decisive game. He is ahead 20-19. He hits a slam, forcing me to run far to the right of the table to defend. I lunge at the ball and make awkward contact.

The ball is headed for the doom of missing the table past the net on the left side. Surprisingly, it hits the top of the metal post of the left-most part of the net, changes direction, and hits Steve's side of the table. He cannot get to the ball.

A long argument ensues over whether the metal post is part of the net. Of course, it is.

By amazing luck, I win the point, stave off defeat, and go on to win the match.

Among the twenty-five students in the accelerated math class, I was, paradoxically and simultaneously, the worst and the best. Eleventh-grade math was precalculus, and twelfth-grade math was calculus. Although my understanding of algebra, geometry, and trigonometry had been perfect, I had trouble grasping the concepts of calculus, which is the differential and integral study of continuous change. I muddled my way through the calculus curriculum by memorizing the procedures and exercises of each module – in each chapter of the textbook – and then spitting them back out on each test. But I did not grasp what was going on. I did not have the inherent understanding. In eleventh-grade precalculus, I got test scores of 92 on average instead of my usual 100. That was still OK. In twelfth-grade calculus, I got nearly the lowest scores on many tests. I was struggling, and the teacher and the other students knew it.

Ironically, in another register, my math and logical thinking ability was beyond all the other students. I got 780 out of 800 on the Math SAT, possibly the highest score in the school. Like in junior high school, I was on the math team that competed against the math teams of other schools. I scored the most points for the entire year of anyone on the team. Then there was the MAA (Mathematical Association of America) Test – I got the highest score of anyone in the school. My score was 78. A different Steven got the second-highest score at 29. All the others scored between 0 and 20. It was sensational. I think it was from that MAA test result that I started (or continued) to believe the – perhaps naïve – narrative that my high intelligence is exceptional. That conviction has kept me alive in many ways over the years. Not the belief itself, but rather the confidence that I had in my ability to apply this self-convincing fictional belief while in the fire or the trenches of specific circumstances, time and time again, when I needed what is usually called a "miracle" to get out of a difficult situation.

At Roslyn High School graduation in June 1972, I was given all three math awards that they had. I was called up to the stage three times at the awards ceremony. And all of this, despite being two years younger than all of them. I was the only student in the graduating class accepted to MIT (the Massachusetts Institute of Technology, America's premier technology university).

> Joan Dziomba swept business education awards with the Dorothy Izzo Memorial and the L.I. Business Education Chairmen's award to an outstanding student. For the top Roslyn score in the Mathematical Association of America Examination, Alan Shapiro was the winner. He also captured first place on the Roslyn Math Team, with second and third places going to Steve Kahn and Rebecca Moore. The Ren-

The Roslyn News, June 15, 1972 (Public Domain, photo by Alan N. Shapiro)

I remember that MAA test vividly. It lasted two hours. I remember sitting there with my pencils and sharpener. I was energetic and "in flow," and I knocked off one question after the next. You could answer a question successfully if you knew, off the bat, the "higher mathematics" solution to the problem. I didn't know those solutions. Alternatively, you could take the long way, figuring it out with painstaking arithmetic and "brute force." And you would have to be very fast – to pull this off for many questions. Probably, the other kids didn't bother to go that route, or it never occurred to them. Not immediately knowing the solution, they skipped to the next question. I knew I could quickly perform the

## CHAPTER 4

"long way" details and make headway at super-speed. Another way of explaining this is to comment on the strange way that my self-confident "transdisciplinary" mind works: I know a little about many things and can do a lot with a bit of knowledge.

For making the major life transition at the end of the twelve years of schooling, the standard procedure was to apply to about five colleges around the end of the calendar year of twelfth grade. Then, in April, you would hear back from them if you were accepted, rejected, or on a waiting list. You had your first-choice school. You had your "safe school," whose standing was below your achievement level. This guaranteed that you were at least going to get in somewhere. After you were officially accepted to college, you could relax. There was nothing more that you needed to do. From early April, the rest of the school year was a walk in the park. You could party. You could have fun and not give a hoot about anything.

I was so stressed out from the suffering of being two years younger – and from all my years of disciplined education – that I decided to sidestep this whole process. MIT offered the option of "Early Decision." I applied for admission in September, at the beginning of my senior year, and was accepted by MIT in December. Since my birthday is in April, I can say that I was admitted to MIT at age fifteen.

The great news of my December acceptance to MIT gave me the gift of six months of freedom with no worries until graduation in June. Yet, in a way, this course of action I had chosen and taken backfired on me. As the senior school year progressed and I approached age sixteen, it became increasingly clear to me that I was more interested in the humanities (literature, philosophy, psychology, history, and politics) than science, math, and engineering. By April, I decided to attend my second-choice college – Cornell University – and not MIT. The Cornell Arts and Sciences division was strong in the humanities. But when April came around, Cornell did not accept me. They put me on a waiting list. Several weeks later, another letter from them arrived in the mail. Unfortunately, they had no place for me in their freshman class. I suspect that the MIT and Cornell Admissions Offices decided together that I would go to MIT since I applied for Early Decision at the prestigious technology university. Having personally questioned my math/science/engineering destiny and having changed my mind about my future since December, I did not fit with their program and was seemingly out of luck.

In my senior year of high school, I had a circle of friends for the first time. I played bridge, table tennis, and some basketball with Steve, Marc, and Andy. We worked on stamp and coin collections. We went to fancy Manhattan hotels to participate in adult contract bridge tournaments. We went to Mets baseball games at Shea Stadium. Andy was my partner in the physics lab. We went to Knicks basketball games at Madison Square Garden. Andy had a driver's license and a car. It was a Corvette Stingray two-door coupé sports car. Its modified V8 engine afforded instant acceleration. We drove to lunch in the middle of the school day in the Stingray. We ate roast beef sandwiches and fried chicken at the newly opened Roy Rogers, Jewish deli food at Andel's appetizing store, and pizza and meatball parmigiana heroes at an Italian joint near the Roslyn LIRR train station. On weekends, I went to the movies frequently with Andy and another Mark.

In the automotive art of Missouri painter Dana Forrester, we see images of Corvettes integrated with classic icons of American pop everyday consumer culture: Corvettes sitting in front of Bob's Big Boy hamburger drive-in and Corvettes parked in front of weathered and layered advertising signs painted on brick walls. The signs are images of Budweiser beer, pinup girls, Speed Shops, Route 66 Service, soda fountain service, and a barbershop. Another Forrester painting shows the exit of the assembly line where the Corvette was produced. In other artworks, we see Chevrolet paired with the Indianapolis Motor Speedway, Tastee-Freez soft ice cream, A&W Root Beer, Fried Chicken Restaurant, filling station, baseball, hot dogs, and apple pie.

"The Humdinger," Dana Forrester watercolors (Reproduced with permission from the artist)

Being two years younger was less of a social-psychological problem for me than in Herricks. One reason for this is that we were a bit older, nearing the end of high school, so the age difference mattered somewhat less. A second reason was the difference in the makeup of the social class of residents of Roslyn as compared to Herricks. Higher incomes made attitudes more liberal and tolerant. Roslyn's ethnicity was also almost entirely Jewish. There was more of an appreciation of high intelligence and academic achievement. I was respected for being close to graduation at only fifteen and then sixteen.

In the homeroom class in twelfth grade, I sat at a table in the back every morning with two others. One was Mike – and the other was Janet. I was deeply in love with Janet. She was the girl of my dreams. And she remained so for a very long time. It took me about twenty years before I finally fell out of love with her and stopped thinking about her. Amazingly enough, at one point in time, she was indeed my girlfriend for two weeks. That happened a year later – while I was an undergraduate student at MIT.

Janet was pretty, and she was highly intelligent. She was one of the smartest individuals in the school. She was not in the accelerated math class. Her favorite subject was English. Her father was a French teacher at Roslyn High School who died suddenly of a heart attack a few years previously. Mike and Janet were friends. Mike was the best male friend of Janet's boyfriend Larry. Mike and Janet talked endlessly between themselves at our homeroom table for three and ignored me. They didn't say more than a few words to me the entire year, nor did I to them. I always listened to their conversation, but it was not especially interesting. I sat there alone and speechless, silently resting my body and mind for a few minutes, waiting to go to my next early morning class.

I was the only male among eight females in my small twelfth-grade Spanish class. I must admit that I enjoyed it. I was fifteen years old to their age seventeen. Being small and young, I was like a child to them, like a eunuch. They adopted me as their "mascot." I was surrounded by female loveliness. They talked about "girl things" all the time, like makeup, jewelry, and boys they liked. And it was also "show and tell." Some of them wore cheerleader costumes. They changed their clothes sometimes in the classroom. Miss Danzig was a fiery, passionate Spanish teacher. It was hard not to have erotic fantasies about her as a dominatrix.

In the second half of twelfth grade, I was in a Social Studies seminar class with Mr. Katz, focusing on the history of Supreme Court decisions. We went through every significant Supreme Court case deliberated over two centuries and held debates on the merits of the attorneys' arguments on each side of the dispute and what the Justices determined and wrote. It was as if we were in law school. It felt engaging but was an information overload of too many details. For the most part, it went in one ear and out the other.

Physics was a fascinating subject. The teacher was great, and the lab exercises were fun. I was the best student in the class. My 96 on the final exam beat out Steve.

In the twelfth-grade English class, we read works of science fiction like Aldous Huxley's *Brave New World*, Ray Bradbury's *Fahrenheit 451* and *The Martian Chronicles*, and Kurt Vonnegut's *Cat's Cradle* and *Slaughterhouse-Five*.[32] I started to love science fiction and understand it as a genre of profound social criticism. We read novels about the Beat Generation and the counterculture, like Jack Kerouac's *The Dharma Bums* and Richard Brautigan's *Trout Fishing in America* and *A Confederate General from Big Sur*.[33] I connected emotionally with the spirit of rebellion against the dominant American culture expressed by these authors. We read classic novels like Charles Dickens's *Oliver Twist*, Herman Melville's *Billy Budd, Sailor*, and Jack London's *The Call of the Wild*.[34] We read poetry by Allen Ginsberg, Sylvia Plath, Emily Dickinson, and Walt Whitman.[35] We read plays by Shakespeare, Henrik Ibsen, Arthur Miller, and George Bernard Shaw.[36] We read the Transcendentalists Ralph Waldo Emerson and Henry David Thoreau.[37] Literature and reading became important to me.

## My First Job in Manhattan

During the summers after the eleventh and twelfth grades, I had my first experience working in the business world in midtown Manhattan. My father gave me a summer job at his engineering firm, the Office of James Ruderman (OJR). The consultancy, called initially the partnership of Ruderman and Severud, was founded in 1927, following a working stay of several years by Mr. Ruderman in post-Revolutionary Moscow. Ruderman was a practical genius at building construction. He also had a keen theoretical or academic interest in structural engineering. He indeed helped the leaders of the Soviet Union with their architectural projects. After returning to America, Ruderman did the structural design of about half of all the skyscrapers erected in New York City from the end of World War II until his death in 1966. It amounted to about 40 million square feet of office space.

James Ruderman died suddenly of a heart attack at age 67. My father and Leo Plofker, who had been junior partners, became 50/50 senior partners in owning and running the firm.

The number of employees rose and fell continuously, anywhere between twenty and sixty, depending upon the state of the American economy, in expansion or recession, which directly affected how much new building construction was going on. There was a lot of hiring and firing happening at OJR. Business was booming the summers that I worked there. The engineers and draftspersons at their desks filled up the fifteenth floor of 515 Madison Avenue between 53$^{rd}$ and 54$^{th}$ Streets. The Art Deco high-rise was also known as the DuMont Building. It was the headquarters of the Dumont Television Network, a longtime rival of CBS and NBC for market preeminence in broadcasting, which ceased operations in 1956.

I had no contact with my father during the workday. We woke early and drove from Roslyn to Manhattan in his green Mercury Montego. We ate breakfast in the coffee shop on Lexington Avenue where my father went daily. Some of his employees and other people whom he knew in the business were there. He sat on a stool at the counter, had his eggs, toast, potatoes, and coffee, and flirted lightly with the waitress who served him regularly. We bought the Daily News or the New York Post at a newspaper stand for a few coins. We looked at the sports section and read about baseball and horse racing. We walked to the office and rode the elevator. Then Dad and I said goodbye for the day. It was 9 AM. I went to the corner of the large open engineering workspace and started my work routine. I was given instructions and assignments by Walter Truhan, the head draftsperson, who was my boss.

At 5:30 PM, I met my father in his inner office. We walked to the garage, climbed into the car, and fought through the rush hour traffic. We listened on the radio for tips on which route to take home from a helicopter guy observing all the cars below. My Dad had a set sequence of radio moments he looked forward to while driving. There was the Bob and Ray comedy routine. There was the Feature Race of the Day from Monmouth Park.

"Hess leads the way in the USA with 101 octane gasoline," proclaimed the audio advertisement of the horse race's sponsor.

We made it back to Roslyn by about 7 PM.

Walter Truhan was an exceptionally nice guy and a close confidante of my father, both personally and in business. He was not Jewish, but he was very much a New Yorker. Although middle-class in income, he was working-class and charismatic in his speech, humor, and attitude. He treated me very well, and we had a good time together. Walter was a father figure for me. Like most men in the office, his politics were right of center. He was a Republican.

It was shocking to encounter all these people with Archie Bunker-like views about life in the rough-and-tumble Manhattan business world. They disliked the "long-haired hippies" of the times. They expressed their fear of non-white minorities and non-white immigrants taking over America. Today, they would be MAGA Trump supporters, begging to be manipulated by and cheering on whatever lies he chose to tell them.

"Vote for that commie radical George McGovern? [the Democratic Party candidate for President in 1972] No way," said Ralph, the amiable "checker" of engineering blueprint details who lived on GOP-dominated Staten Island. "He wants to give away a thousand dollars a year to those lazy bums who don't want to work! Nixon is the one. He's an honest, upstanding guy. Four more years!"

"Screw those college students with their flag-burning," said his amiable co-worker Stuart to me. "Get the hell out of our country! Love it or leave it!"

I was an "office boy" with various tasks to perform.

There were massive metal filing cabinets with hundreds of long, thin drawers containing large-sized drawings or "tracings" of every floor of every past and present office building project of the firm. I was responsible for keeping the classification and contents of the drawers organized. Engineers and draftspersons asked for specific tracings to work on. When they were finished, the tracings needed to be rolled up properly and returned to the right place.

In 1971, the monopoly technology company IBM introduced the System/370 with the 3270 computer terminal. I learned to type code and data onto punch cards and how to execute a computer program with the card deck.

I learned to operate the manual telephone switchboard exchange. I "relieved" (took the place of) the black female operator when she went to lunch or was out for the day.

I speak with the incoming caller: "Office of James Ruderman. How may I help you?"

I ring up the intended call recipient on their internal phone extension. If they assent to speak with the caller, I establish the connection by pulling up an electrical cord and plugging its end into a socket hole. I turn a black knob at the base of the switchboard to the right to speak with both parties and politely introduce their verbal exchange by saying their names.

I also give individuals inside the firm outside lines upon request.

The highlight of the job was being a messenger. I delivered or picked up tracings or other documents to/from architectural and real estate owner offices with which we did business on projects. I left the office and was on the Manhattan streets in the summer. I could talk with or look at people. I could take extra time and say that the subway got delayed. I could get a snack to eat. I could wander around shops. Occasionally, my father took me to construction sites. I saw the erecting of the building in progress. I saw how concrete is poured.

I liked the 10:30 AM coffee break every day. A food cart came to a hall near the elevators. The vendor rang a loud bell. They had good cakes wrapped in thin plastic cling film.

## CHAPTER 4

During the summer between twelfth grade and my freshman year of college at MIT, I suddenly "shot up" seven inches (eighteen centimeters) in height. It was an inherited genetic pattern of my physical growth. The transformation happened in just a few weeks. Over several years, I gradually grew from four feet nine to five feet three. Suddenly, the big stretch came. At five feet ten, I was now an average normal height for a man.

I was ready to travel to Cambridge, Massachusetts. I was no longer a shrimp.

# Chapter 5

# Massachusetts Institute of Technology

"This is Dad from beyond the grave, Alan. I've read what you've written so far. You write well. You've managed to avoid the accusation I was going to make that you indulge in self-pity and blame the world for troubles of your own making. I commend you for that. But now you're going to write about your two years at MIT? I hope you think this through carefully before putting your fingers on the keys. You're not going to criticize MIT, are you? That's a fine university. It's America's most outstanding technology university and renowned throughout the world. If you were unhappy there, you should squarely blame yourself. You can't always claim a monopoly on high moral grounds. Remember the meaning of your *bar mitzvah*. You are responsible for your actions and what happens in your life. Stop blaming others and be a *Mensch*."

"OK, Dad. In recounting what I experienced as an undergraduate at MIT, I do not intend to criticize MIT. I ultimately found something great at MIT that was very valuable to me: their excellent political science department. I will make sure to give some serious praise to MIT. Overall, MIT was not the right university for me. I already hinted at that situation in my anecdote about the benefits and detriments ensuing from my having applied to MIT 'Early Decision.' The problem is related to the metamorphosis of my intellectual interests from engineering to the humanities. But that's just me. MIT was certainly the right learning institution for almost all the undergraduate students who were there. I was the odd duck. My circumstances were unhappy. It was not MIT's fault."

At the end of twelfth grade, I wanted to go to Cornell for humanities and not to MIT for engineering. After two years at MIT, Cornell finally accepted me, and I switched to Cornell as a transfer student. However, I got a lot of valuable education during my two years at MIT – both in political science and other scientific disciplines – that I would not have gotten if I had spent all four undergraduate years at Cornell. I had educational experiences at Cornell that were unavailable at MIT – two years at each university turned out to be the best of both worlds.

During four semesters at MIT, not only did I get straight A's (although the first two semesters were on a pass-fail basis), but I accumulated 216 credits of the 360 credits required for graduation. I was close to getting my Bachelor of Science degree.

## My First Year at MIT

I was sixteen years old and – other than the four summers at the camp in New Hampshire – away from home for the first time. My first semester at MIT got off to a bad start when I made a critical error in choosing a dormitory roommate, repeating the psychological pattern of my earlier "Early Decision" mistake. In both cases, I sought an easy way to "get something over with." My unfortunate tendency was to want to get past an uncertainty or seek "premature closure quickly." I got stuck with a terrible roommate named Eric. He was also from Long Island. We met on the first day of the freshman orientation week. Some lucky students were going to get single rooms in dormitories. Most of us needed to find a roommate to share a double room. Eric seemed amiable enough when we first met, so we agreed to live together.

However, after the first day of the orientation, I had no further contact with Eric for the rest of the week. But we already signed up to be roommates, which could not be changed. Eric joined a group of friends planning to study electrical engineering like him. He and his new friends started smoking pot every day, which I did not do. A large percentage of the MIT undergraduates living in dormitories smoked a lot of marijuana and hashish and were constantly getting high.

Once we moved into our shared room, Eric and I didn't connect. We stopped speaking to each other. He took a dislike to me – for reasons I did not understand.

Eric claimed that I snored at night and became very angry about it. One time, I was deep asleep. He approached my single bed and screamed, "Shut up!" directly into my ear.

Our desks were adjacent, but I had to study in the library. While he was stoned "out of his mind" and working on his math and engineering assignments, he played rock music at screeching maximum volume by the progressive group "Yes":

*Close to the Edge*, by Yes

Only to find out the master's name
Down at the end, round by the corner
Close to the edge, just by a river

Seasons will pass you by
I get up, I get down
Now that it's all over and done,
Now that you find, now that you're whole

I get up, I get down
I get up, I get down (Atlantic label)

I felt like I had "found out the master's name," at least the name of the master of our dormitory room in Burton House. His name was Eric.

The first semester at MIT was all required courses except for half a course out of five. I had to take calculus, physics, chemistry, and economics. These were all lecture courses in large lecture halls with hundreds of students. Attendance was not mandatory. I went to the lectures slightly more than half the time. Much of the work was self-study and preparing yourself for exams. I was a good test taker and did well in that system.

I still did not understand calculus, yet I passed the tests by monkey-like memorization of the steps to complete the problems. It was almost entirely a repetition of the twelfth-grade calculus class at Roslyn High School.

The physics lecture course almost entirely repeated what I had learned in twelfth grade at Roslyn High School. The material in the chemistry lecture course was new, but I found it to be very dull. "My bad. I admit it, Dad."

The lectures in the Economic Principles course were given by Professor Richard Eckaus.[38] We read the classic introductory economics textbook by MIT Professor Paul A. Samuelson.[39] Samuelson later married Eckaus' ex-wife. From Samuelson's book and Eckaus' lectures, I absorbed some of the vocabulary and concepts of what leftists would call "bourgeois economics." I was curious about Karl Marx. The references to Marxian economic theory by Samuelson and Eckaus were scarce, and they did not understand Marx well. As an alternative to Eckaus and Samuelson, I preferred to read the more liberal Harvard economist John Kenneth Galbraith, such as his books *The Affluent Society* and *The New Industrial State*.[40] Galbraith seemed much more attuned to the inherent social injustices of the American and capitalist economic systems.

All freshmen had to take one humanities course. We had a choice of one among several offers. I chose the political science offering called "Modernization of Society." Political science is, strictly speaking, not a humanities subject. Since that poli sci course had a huge number of registered students, each of us was arbitrarily assigned to one of many smaller lecture modules with a specific professor. I was assigned to the module taught by Professor Myron Weiner.[41] His special area was the history, politics, and social problems of the Indian subcontinent. I learned some things about India and Pakistan, but I was too immature to appreciate the humanity behind Weiner's commitment to an "objective" academic presentation style. I thought, naively, that he was boring. "My bad again, Dad."

CHAPTER 5

The phrase that I heard again and again from my fellow MIT students in my dormitory – from both freshman and upperclassmen – was their emphatic "humanities is bullshit." It was their mantra. One of them said it to me in the cafeteria. Another one said it to me while drinking a beer. Another one said it in an elevator. I heard them saying it to each other. Almost all of them chose civil, mechanical, or electrical engineering, mathematics or physics, or biology or chemistry (pre-med) as their "major" course of study. Only one time during my freshman year did one of these MIT undergraduates say to me something like: "You know, Alan. They are wrong when they say that 'humanities is bullshit'. Humanities is something very important and valuable." The person who said that was my friend John Asinari, a pre-med student and lacrosse player from nearby Arlington, Massachusetts, who envisaged a humanitarian medical career. John died tragically at age twenty in an incident of murdering brutal violence. I shall return to that later.

Besides keeping up with my five courses and steering clear of Eric's booming stereo speakers, my principal activity during the Fall of 1972 was working for the George McGovern for President campaign. McGovern was a close ally of the assassinated Robert F. Kennedy. I met some nice and intelligent people among the campaign volunteers but made no lasting friendships. McGovern's national campaign was sadly in deep trouble. It appeared that Massachusetts would be the only state whose electoral college votes he would win. Since the Bay State was in the bag, I was put on a bus a week before Election Day and sent to Lewiston, Maine, to do campaign work there. The hope was that maybe McGovern had a chance to win the Pine Tree State. I went door-to-door to houses and apartments, handed out campaign literature, and talked with potential voters about issues. At night, I slept in a sleeping bag on the floor of a storefront on a street where one could see that many businesses had gone bankrupt. We ate bologna, tuna fish, and peanut butter and jelly sandwiches.

On Tuesday evening, November 7, 1972, I rode with several people in a station wagon on the highway from Lewiston back to Cambridge. We stopped for dinner at a hamburger diner near Portsmouth, Maine. President Richard Nixon's victory was such a landslide that the overall result was already known by 8 PM. We listened to the news on the radio. It was projected that Nixon would get 61 percent of the popular vote to McGovern's 38 percent. I felt devastated while eating my burger and fries. At that moment, I felt deeply that something was wrong with America. McGovern was the good guy. Nixon was a crook. Nixon was a Vietnam War monger. Nixon and the Republicans did not care about poor people. Nixon was involved in the "dirty tricks" break-in at the Democratic National Committee headquarters at the Watergate Office Building in Washington, D.C. Stories about the legal entanglements of Watergate appeared in the newspapers since June 1972. Still, only dedicated news readers paid attention to them.

The percentage of female students at MIT was only about 10 percent. The few female students were like exotic animals. I had no contact with girls. I had a couple of male friends but almost no social life. I went home to Roslyn for Christmas vacation at the end of the Fall semester. One evening just before Christmas, my twelfth-grade friend Steve hosted a class reunion party in his basement. About fifteen of our "high-achieving" male classmates from the Roslyn High School Class of 1972 were there. They greeted me warmly. They were amazed that I was now slightly less than six feet tall instead of slightly more than five feet tall.

In the company of these Roslynites, I started thinking about Janet from my senior year homeroom class in high school, with whom I was in love. I drank a couple of glasses of wine and confessed to Paul, who knew both Janet and her circle of friends well, my feelings for her. Paul reacted sympathetically. He said that I should call Janet. He strangely encouraged me to act on my secret crush. He reported that Janet broke up with her former high school boyfriend, Larry, a while ago and was now single. He believed that she would be happy to hear from me.

I called Janet on the phone the next day, and we met. She was staying at her mother's house during the Christmas break. It was only 2.5 miles (4 kilometers) from my parents' house. Janet was a student at an elite university in Connecticut. We spent every day together for the next two weeks. The girl of my dreams was my girlfriend for a fortnight. I was sixteen years old to her age eighteen. I couldn't drive a car yet, but she had a driver's license and her mother's car. She picked me up every day. We discussed "God and the world" or "everything under the sun." We went for walks and listened to music. We went one time on a day trip to Manhattan. We had dinner with my parents in a lobster restaurant in Great Neck. My parents liked her. Janet and I kissed only once, like Yuri Gagarin's one orbit around the Earth. Or maybe more like Chuck Yeager (played by Sam Shepard) in *The Right Stuff*, climbing in his Lockheed NF-104 aircraft to superhigh altitude and touching the edge of outer space for one short moment before crashing down yet surviving. I think it was an achievement and triumph that I had those two weeks with her.

At the end of the two weeks, Janet broke up with me. Her explanation, which I believe was truthful, was that she decided to give a chance in love to Mike, the boy who sat with us at our table in the back of the twelfth-grade homeroom class. Mike had also been in love with Janet for a long time. He recently declared his love to her. Mike was a freshman at the same college in Connecticut as Janet. He was a close friend of her ex-boyfriend Larry and stayed quiet about his feelings for Janet until the original couple broke up. Janet said she liked me more than

CHAPTER 5

she liked Mike, but he should now be given his chance in the interest of fairness because he was "first in line" ahead of me.

I was deeply hurt. I was still in love with her and would remain so for a long time, but I also developed anger towards her. I was too young to know how to manage those negative emotions. "My bad. I admit it, Dad. Once again."

During the inter-semester break in January 1973, I participated in an interview and research project sponsored by the MIT political science department and the Sloan School of Management. At the 1968 Democratic National Convention in Chicago, Hubert Humphrey was selected by the party establishment as the nominee for president without having democratically won even a single state primary. Before the 1972 presidential campaign, reforms were introduced to make the nomination more decided by democratic primary elections and less by party bosses – and to have more delegates to the Convention representative of African Americans, women, and young people. My participation in the project involved conducting and analyzing many live interviews with individuals in the New York City area who were delegates to the 1972 Convention.

My second semester at MIT in the Spring of 1973 was not especially eventful. There was more calculus, more physics, and more Economic Principles. There was no more required chemistry class. There was another semester of the introductory political science course "Modernization of Society." I was allowed one purely elective course. I took the "American Political Process" seminar with Professor Louis B. Menand.[42] Menand was an impassioned thinker and speaker. In addition to talking about American politics and constitutional law, Menand gave great explications of the classic texts of liberal political philosophy and social contract theory of John Locke, Thomas Hobbes, and Jean-Jacques Rousseau.

The dormitory managers allowed me to change roommates and escape from my nemesis, Eric. For my second semester at MIT, I had a better roommate. He was a third-year student majoring in urban studies and planning. His beloved hobby was square dancing.

One afternoon in March 1973, I opened my mailbox in the lobby of my MIT dormitory. There was a handwritten letter from Janet. She wrote that she was in love with me and wanted to meet with me soon, possibly in Roslyn during Spring Break, which was a few weeks in the future. Janet recounted that her experiment to be together with Mike in February had only lasted a few hours. She realized quickly that she was not attracted to him romantically and wanted to be with me. The letter was a fantastic surprise and made me happy. It gave me hope for happiness. I wrote back to her, and she wrote me a second letter. We arranged to spend time together again in Roslyn and see where things go. But there was a practical problem. She would not have a car available during the Spring Break. How we would travel between her mother's and my parents' houses was unclear.

I felt frustrated that I did not yet have a driver's license. In New York State, the age for getting your license was eighteen. But you could get it at seventeen if you took the "Driver's Education" course in high school when you were sixteen. Although I was now seventeen, I was stuck waiting until eighteen because I had been too young in high school to take that course.

Next came one of the stupidest and most self-destructive things I ever did. When I arrived in Roslyn for the Spring Break in mid-April, I was ready to call Janet on the phone. But I didn't want to call her from my parents' house. I went to the house of the parents of my high school friend Steve and decided to call Janet from there. Steve was, in fact, a very competitive guy with a hardened psychological character and little empathy. I was already questioning, in my mind, if I wanted to remain friends with him. But I made the mistake of telling him I was angry at Janet after she broke up with me in January.

Steve lived in a big house. His parents were not at home. I entered his parents' bedroom, picked up the rotary phone receiver, and dialed Janet's Roslyn phone number. She answered, and we started to talk. Steve was downstairs in the kitchen, where there was another phone extension. He heard that I was talking on the phone upstairs. He picked up the receiver of the kitchen phone. He asked who I was talking with. I said it was Janet. He then exclaimed: "But you told me that you hate her!" Janet hung up instantly. It was a disaster. It did not occur to me to call Janet back. I was overwhelmed by the situation. There was no way to undo what had just happened. I did not try to speak with Steve about what he did. Nor what I did. I knew that I would never speak with her again the instant she hung up. I also knew that my friendship with Steve was finished.

During the summer of 1973, I worked another six weeks at the Office of James Ruderman. In the early summer, I went hitchhiking around the northeastern United States for a couple of weeks with my MIT friend Dave, the best friend of John Asinari, who, eighteen months later, would meet his violent, tragic end. Hitchhiking in America was eminently foolish. We were fortunate that nothing bad happened to us and that we were not victims of a violent incident. On June 10, eighty thousand others and I attended the concert of the Grateful Dead, Wet Willie, and the Allman Brothers Band at RFK Stadium in Washington, D.C.

*Ramblin' Man* (1973), by the Allman Brothers Band

Lord, I was born a rambling man,
Trying to make a living and doing the best I can
And when it's time for leaving, I hope you'll understand,
That I was born a rambling man

Well, my father was a gambler down in Georgia
And he wound up on the wrong end of a gun
And I was born in the back seat of a Greyhound bus
Rollin' down Highway Forty-One (Capricorn)

## My Second Year at MIT

When I returned to MIT in September 1973 for my second year, I registered officially as a political science major. I still had to take one more required natural science course and two courses in "Psychology and Brain Science." Other than that, I was free to take elective courses. For the final mandatory science course, I took "Environmental Ecology." The teacher worked at the Woods Hole Oceanographic Institution and told stories about fishery management.

Part of my growing skepticism about the morality of having the identity of being a student at MIT was concern over MIT's role in the American military-industrial complex and the Vietnam War. One of my elective courses in my third semester was "Introduction to Film Making" with Edward Pincus. Pincus and Richard Leacock, a celebrated experimental documentary filmmaker, founded the MIT Film School.[43] In the class, Pincus had us shoot and edit short movies with Super-8-millimeter film. We learned about *cinéma verité* as a documentary genre and aesthetic philosophy. We watched the films of Frederick Wiseman: his witnessing to real-time lived experiences in hospitals, basic training in the military, and youths in trouble in a juvenile court in Memphis. Pincus, however, emphasized the subjectivity of filmmaking. He wanted us to get personal in both form and content. He believed the artist should explicitly acknowledge himself rather than hide behind a positing of objectivity.

In his unfinished film of 1969, *The November Actions*, Richard Leacock documented the anti-Vietnam War student movement at MIT. The movement was not only opposed to Johnson and Nixon's war in Southeast Asia but also acted against open and secret military research done at MIT financed by the Pentagon and large companies with military contracts. The U.S. government notoriously deployed horrendous weapons in Vietnam, like the flammable liquid anti-personnel chemical Napalm and the "herbicidal warfare" defoliant Agent Orange. It was believed that classified military research was going on in the MIT Instrumentation Lab (later renamed the Draper Laboratory), which worked on advanced technologies for national security. The lab had developed the Polaris submarine-launched Intercontinental Ballistic Missiles (ICBMs) guidance system for nuclear weapons delivery since 1957.

One of my two elective poli sci courses in that third semester at MIT was "International Relations." This was a popular seminar attended by about thirty-five students. The female professor had a celebratory perspective on American diplomacy and its role in the interactions of nation-states and world affairs akin to Henry Kissinger's worldview. I also wandered into Noam Chomsky's lectures on the complicity of intellectuals with wealth and power and his critique of U.S. foreign policy.[44] The two venues stood in marked contrast to each other.

A tenth-anniversary article about the Kennedy (JFK) assassination that I wrote and published in the MIT independent community weekly newspaper "Thursday" on November 29, 1973. (Public Domain, Alan N. Shapiro private collection)

The number of African American students at MIT was extremely small. My second elective political science course in the third semester was "Black American Politics." I was the only white student in the class. The brilliant Professor Lorenzo Morris taught the seminar.[45] Morris was a young intellectual, only ten years older than me. He became America's leading scholar of the history of black politics.

It was a small group, and we had amazing discussions. Lorenzo talked eruditely about the ideas, social science, and political commitments of W.E.B. Du Bois.[46] The following semester, I did another seminar with Lorenzo Morris called "Urban Politics." Lorenzo also suggested that I read *The Structure of Scientific Revolutions* by Thomas S. Kuhn.[47] Kuhn's landmark work gave me insights into the institutional contexts of scientific research and ignited my interest in the philosophy and history of science.

I was close to eighteen years old. I had read many books for my general education. That was good. I read the books assigned in school. But I had still not thought for myself. I had not yet developed my worldview in a focused way. I had no distinct clarity. I had not yet read any serious books I had consciously chosen, on my own, to read. I had not reflected on the meaning of life. I did not know what I thought about God and religion. I did not know what my political views were. I knew I was left-liberal but did not know precisely where I stood. It was time to choose books to read and engage in serious dialog with authors who might help me. I was on my way to the existentialism of Jean-Paul Sartre and Albert Camus.[48]

The British analytical philosopher Bertrand Russell was the first author whose books I chose to read, whose ideas I grappled with seriously, and who influenced me.[49] Much later, I read Ludwig Wittgenstein, a student of Russell's who criticized Russell convincingly as too rationalist.[50] I then saw Russell's limitations. My period of reading Russell at age seventeen was short-lived yet consequential. I read about five of his books. The main event was my reading of *Why I Am Not a Christian*.[51] In that book, Russell asserts that the power of religion is rooted in fear. He deconstructs with logical reasoning the principal traditional philosophical arguments for God's existence: the first case argument, the natural law argument, the design argument, and the moral argument. Russell further maintains that religion stands in the way of humanity's progress and society's betterment. The world needs more kindness than Christianity offers. The world needs scientific knowledge and the spirit of open-minded inquiry.

The book had a profound effect on me. I decided that I was an atheist, or at least an agnostic. I have not stayed consistent all my life in this view. There have been times when I believed that I saw an apparition of the divine in the sense of Jewish mysticism or Kabbalah. I have also lived phases of involvement with Buddhist spirituality. But most of my life, I have been an atheist, and I stand my ground on that. Sometimes, I have said that what truly makes God great is his paradoxical and special quality of non-existence.

Bertrand Russell was also a socialist and a historian of philosophy. I read his lengthy tome, *A History of Western Philosophy*.[52] I read his books *The Conquest of*

*Happiness, In Praise of Idleness,* and *War Crimes in Vietnam.*[53] I read some of his three-volume autobiography.[54]

George Orwell influenced me a lot. Beyond his critiques of totalitarianism in *1984* and *Animal Farm*, I learned about his commitments to democratic socialism and even to anarchism by reading *The Road to Wigan Pier, Homage to Catalonia,* and his famous collected essays.[55]

Existentialism was in vogue among philosophically oriented college students in the 1970s. I started to acquire anthologies of writings of "the existentialists." I was introduced to Fyodor Dostoevsky, Franz Kafka, Søren Kierkegaard, Karl Jaspers, and Friedrich Nietzsche.[56]

After all those political science courses, I decided to delve deeper into the humanities. In my fourth and final semester at MIT, I registered for the introductory philosophy course "Problems of Philosophy" and the introductory literature course "Approaches to Literature."

The literature course was pretty good. We read William Shakespeare, Herman Melville, Joseph Conrad, James Joyce, Flannery O'Connor, Virginia Woolf, and Chinua Achebe.[57] This was all great for my general education. The professor made exciting interpretations. However, he spent much time explaining and justifying why it was valuable for science and engineering students like "us" to read literary works. He was well-intentioned, but it made me sense that literature studies at MIT would never get deeper or do justice to the material.

The philosophy seminar was worthwhile yet disappointing. We read short excerpts from classical texts by great Western philosophers like René Descartes, Immanuel Kant, David Hume, George Berkeley, Baruch Spinoza, Gottfried Wilhelm Leibniz, Thomas Aquinas, Saint Augustine, and Willard V.O. Quine.[58] These readings were valuable for my first introduction to philosophy. There was little to no discussion, even though the class was small. The professor did all the talking. He was dry and boring. It seemed that for academia, a discourse was only valid as legitimate philosophy if it was argued with rigorous logic, rationality, and the so-called detached, objective intellect. There was nothing about life. Nothing about experience. Nothing about embodiment or emotions. Nothing about alienation, contextual ethics, or the interspaces between persons. It was not existential. There was not going to be any of that at MIT.

At the start of my second year at MIT, I moved to a different dormitory where I had a single room. The new dorm, MacGregor House, was at one end of the vast MIT campus. The political science building and the Sloan School of Management, where I had many of my classes, were at the far end. It took me about thirty minutes on Memorial Drive along the Charles River to walk each way between the dormitory and the building where my classes were and where the poli sci library was located.

At seventeen years of age, my sexual drive was beginning to overwhelm me. This was already very late in the day. I had no contact with girls. I was getting lonely. I started to look at soft pornographic magazines and masturbate. One night, I met a young woman at a fraternity party. She was a nursing student at a college across the river in Boston. I rarely drink alcohol, and I don't like beer, but that night, I got drunk on beer drinking. The woman and I went to my room and made out for a couple of hours. She gave me her phone number and wanted me to call her. But I never did.

In April 1974, a letter from the Cornell University Admissions Office arrived in my dormitory mailbox. Cornell Arts and Sciences accepted me as a transfer student. I could start there in September. I would go to Ithaca, New York. My high school friend Marc was at Cornell. He expressed his willingness to be my roommate in a semi-off-campus communal house. I visited Ithaca and met a couple of others in Marc's circle of friends. Marc was a year older than me in age and a year younger than me in school. He went to Great Neck South High School, but I met him through inter-high-school table tennis and contract bridge.

Marc Silver, my lifelong best friend (Photo by Alan N. Shapiro)

I didn't know exactly what I was going to study at Cornell. During my last months at MIT, I befriended Bill, a fourth-year premed student living on my dormitory floor. Bill had logistically done the opposite of what I was doing. He transferred from Cornell to MIT. One evening, we sat having dinner in the cafeteria on the ground floor of the dormitory. We had a serious personal discussion. We then walked out into the courtyard. It was warm weather. It was one of those moments when you feel like fate is showing you the way.

"Alan, when you get to Cornell, take the course in European Intellectual History with Dominick LaCapra.[59] It will be perfect for you. People think that LaCapra's lectures are great."

On May 17, 1974, I walked into the MIT Registrar's Office and withdrew. When I went home to Roslyn and told my parents I had decided to transfer from MIT to Cornell, a significant argument broke out. They told me that I was making a big mistake. I don't know the deeper reason for their apprehension and opposition. They already knew that I was studying political science, not engineering. My mother's rationale seemed incredibly banal. She claimed that dividing my undergraduate years between two universities would look bad on my resumé, which future potential employers will judge.

In March 1975, ten months after I left MIT, came the news of the horrific postscript to my time there. Four criminal youths brutally murdered my friend John Asinari. John and his friend Robert, another MIT student whom I knew, were kidnapped in a car while hitchhiking from Boston back to Cambridge late at night. The assailants beat and shot John and Robert. They stabbed them multiple times. When John got out of the car and tried to flee, one of the attackers struck him on the head with a tire iron. John died a few hours later from a crushed skull and maceration of the brain. Robert survived by "playing dead." It was an act of hideous sadism and indifference to life. The incident contributed to my desire to leave America.

CHAPTER 6

# Cornell University Arts and Sciences

Now I was eighteen years old, freed from the establishment earnestness of MIT, freed from MIT's half-hearted effort to deal with the humanities, half-estranged from my parents, eager to experience a more libertine life at Cornell, intellectually serious, yet in many ways a lost soul. I felt estranged from religion yet uncertain about an alternative meaning of life.

I was a leftist critic of American society and capitalism, yet I acknowledged the validity of my parents' criticism of the leftist criticism of capitalism. I fought with my parents, yet I am grateful to them because their influence moderated my views. During the Cold War, achieving a morally coherent political worldview was difficult. The totalitarianism of state socialism, as demonstrated in practice by the Soviet Union and its satellite regimes, was something to be opposed. Yet, seeing the wrong-headedness of state socialism should not become a justification for being uncritical of capitalism or America, as was the excuse or pseudo-logic pronounced by my parents. Many intellectuals feel strongly antagonistic to capitalism (criticized by the Left) or socialism (criticized by the Right). I take both "critiques" of the antithetical economic systems (and their antithetical critiques) seriously. I have a balanced view of capitalism, seeing both the good and the bad. I have a balanced view of socialism, seeing both its merits and flaws. Above all, I search for a democratic, decentralized, or libertarian socialism. The year 1968, with its uprisings from Columbia University in New York City to Paris (rebellion against capitalism) to Prague (rebellion against state socialism), was the hope of a society of socialism and freedom.

I spent most of the summer of 1974 painting houses and doing odd jobs with a group of friends from Great Neck who formed a makeshift company to make some money. We placed an ad in a local Long Island newspaper and acquired some clients. I was a mediocre house painter (although later in life, I would improve somewhat). I am grateful to my late teenage summer friends for overcoming my slowness and sloppiness.

I had my driver's license and a used silver Buick Regal that my father gave me after he bought his new model. With the gas-guzzling car, I went on a two-week trip through Canada with my fellow house painter and friend Dave. We saw Toronto, Montreal, Quebec City, and Trois-Rivières. We went camping in splendid countryside with sleeping bags and a tent. I was reading *The Trial* and short stories by Franz Kafka and *The Stranger* by Albert Camus in English translation.[60] We met two French-Canadian girls reading Camus – in French. We were together but had a language barrier. We could not speak each other's language.

Dave was a very sensitive guy from Great Neck who had psychological conundrums with his parents, which I did not understand, and which were more profound and more complex than my conflicts with my parents. He was reading existentialism and was a lost soul like me, not knowing what to do with his life. A few years later, in his twenties, he committed suicide. He stabbed himself with a knife from his deceased father's prized historical knife collection.

In late August, I went to Ithaca two weeks before registration and the start of classes for the Fall semester of my junior year of college. It was gorgeous, hot summer weather. I met a girl with whom I spent some time together, although I lost track of her a few weeks later. There is much alluring nature near the Cornell campus. Every day, I went with a towel, food, and water into the gorge to an idyllic lagoon next to a waterfall on the Cascadilla Gorge Trail.

At that spot, with the roaring noise and the flowing majestic visual beauty of the pouring water as background, I meticulously read Camus' *The Myth of Sisyphus* ten times.[61] I wrestled in my mind, and in dialogue with the words of the long-dead Camus, with the question of the meaning of life for a non-believer. At the end of the long reading session and meditation, I felt I had come up with an answer. I was never suicidal in the desperate or psychiatric sense. I was lost. I was not depressed. I thought deeply about suicide or the meaning of life in a philosophical or even rational-logical sense. Camus saved me, not from death, but from being alone without self-confidence. He saved me, perhaps, from nihilism. Regarding God, Camus provided a reasonable agnosticism. He wrote of creativity, the struggle to create meaning in the "desert" of no given meaning, to find a purpose in one's life.

In *The Myth of Sisyphus*, published in Paris in 1942 during the Second World War, Albert Camus wrote the following sentences at the very beginning of the book:

> There is but one truly serious philosophical problem and that is suicide. Judging whether life is or is not worth living amounts to answering the fundamental

question of philosophy. All the rest – if the world has three dimensions, whether the mind has nine or twelve categories – comes afterwards. These are games; one must first answer.[62]

This famous opening passage has often been misread or misinterpreted. *The Myth of Sisyphus* is not a book of philosophy. Camus is not a professional philosopher. Camus was a great thinker who thought deeply about life. Here and in the following pages, Camus makes a brief foray into philosophy – to find out, through reasoned experimentation, his position on one fundamental single question: the meaning of life. Camus undertakes this experiment not to return to the question repeatedly, like on an endless Wheel of Samsara (as described by Buddhism) or as a repetition compulsion (as diagnosed by psychoanalysis), but to *answer the question*, leave philosophy, and move on. Camus asks and answers the question, draws his conclusions, and then lives the consequences of his reflection. Asked and answered, as the trial lawyers say. As a "transdisciplinary thinker," Camus even expresses a certain disrespect for philosophy as a mono-discipline.

Do you want to live? Camus answered in the affirmative. Yes. The answer is that life, indeed, is worth living. The sense of "the absurd" that gave rise to the question and doubt is not a static condition. The absurd is a dynamic, a relationship, a gap, a cleft – between my aspiration for a good life and the frustration of the existing social-existential order. This dynamic is the groundswell of the most essential human quality: creativity.

Nietzsche writes in *Thus Spoke Zarathustra*: "The child is innocence and forgetfulness, a new beginning, a sport, a self-propelling wheel, a first motion, a sacred Yes."[63] Creativity involves a connection with childhood beyond the separation of play and work.

Camus links suicide and the sense of the absurd, or more simply, the absurd:

> What then is that incalculable feeling that deprives the mind of the sleep necessary to life? Thrown suddenly into a universe divested of illusions and lights, man feels an alien, a stranger. His exile is without remedy since he is deprived of the memory of a lost home or the hope of a promised land. This divorce between man and his life, between the actor and his setting, is properly the feeling of absurdity.[64]

The alien. The stranger. Exile. Divorce. An actor out alone on the stage, performing. Camus binds his "existentialist Marxism" directly to the difficult daily challenges of the proletarian experience. Man or woman is condemned to be free, says Jean-Paul Sartre. Jean Baudrillard says we are now living a parody of that freedom in the consumer culture, forcing us to choose between products with simulated differences. You must choose. Coke or Pepsi. McDonalds or Burger

King. Capitalism offers the worker myriad diversions to get him to forget his giving up his autonomy to the factory or office mechanical job. Camus was a critic of capitalism, and especially of the way that it forces us to live:

> A man wants to earn money to be happy. His whole effort and the best of his life are devoted to earning that money. Happiness is forgotten; the means are taken for the end.[65]

But then comes the beginning of consciousness or awareness:

> The stage sets collapse. Rising, streetcar, four hours in the office or the factory, meal, streetcar, four hours of work, meal, sleep, and Monday Tuesday Wednesday Thursday Friday and Saturday according to the same rhythm – this path is easily followed most of the time. But one day the 'why' arises and everything begins in that weariness tinged with amazement.[66]

Camus concludes logically and pragmatically that suicide is not a solution to the absurd. The absurd does not dictate death. The absurd emerges from the confrontation between the human striving for reasonableness and "the unreasonable silence of the world."

> This world itself is not reasonable; that is all that can be said. But what is absurd is the confrontation of this irrational and wild longing for clarity whose call echoes in the human heart. The absurd depends as much on man as on the world [...] It is all that links them together.[67]

From the moment absurdity is recognized, it becomes a passion, the seed of a seduction. The absurd is born of the desert. Absurdity arises from a comparison or tension. Absurdity is a water source, an oasis in the desert. It appears at first to be a negative, but it is a double-positive of consciousness and rebellion, which are Camus' two basic principles.

The sense of the absurd grows into the historical situation of man in revolt – *L'homme révolté* – like Spartacus, who led the rebellion of the slaves in ancient Rome, one of Camus' primary examples, a story engraved into our minds by Stanley Kubrick's 1960 film *Spartacus*. Revolt or rebellion is the subject of Camus' book *The Rebel*.[68] Camus criticized the orthodox Marxism popular among French leftist intellectuals at the height of the Cold War.

One must abide in the existential and historical situation of the rebel without giving in to the nihilistic temptations of suicide or murder, which are amoral equivalents. The Marxist revolutionary, contrary to the rebel, is willing to commit murder or to hail the "dictatorship of the proletariat" that will "wither away" on the assumption that "the ends justify the means."

Albert Camus, year 1957
(United Press International, Wikipedia Public Domain)

Sisyphus and his rock.

"The gods had condemned Sisyphus to ceaselessly rolling a rock to the top of a mountain, whence the stone would fall back of its own weight. They believed there is no more dreadful punishment than futile and hopeless labor [...] At the very end of his long effort measured by skyless space and time without depth, the purpose is achieved. Then Sisyphus watches the stone rush down in a few moments toward that lower world, whence he must push it up again toward the summit. He goes back down to the plain. It is during that return, that pause, that Sisyphus interests me. That hour is like a breathing space that returns as surely as his suffering, the hour of consciousness."[69]

The figure of Sisyphus represents neither acceptance of the proletarian condition nor simplistic rejection of it. One must abide in the proletarian condition, as in the human condition, with awareness, to transform this condition into something better, creativity, and, later, a better society. The "cultural theory" must emerge slowly and immanently from the experience, from deepening familiarity with the condition. "The struggle itself toward the heights is enough to fill a man's heart. One must imagine Sisyphus happy."[70]

Sisyphus stares down the barrel of the absurd. He comes to terms with the quandary of the human condition, including facing death. The ethics of "the rebel" involves staying faithful to the original act of challenging the unjust situation into which "the gods" have thrown me. I must embody these twin values of creation and revolt in my choices in the present.

## Dominick LaCapra and European Intellectual History

My two years at Cornell were the heart and soul of my lifetime's education. The two semesters of nineteenth- and twentieth-century European Intellectual History lectures by Professor Dominick LaCapra were a fantastic revelation for me. I also did a graduate-level seminar with him in literary theory where we read classics in that field like *Mimesis: The Representation of Reality in Western Literature* by Erich Auerbach, *Anatomy of Criticism* by Northrup Frye, *Theory of the Novel* by György Lukács, and *Writing Degree Zero* by Roland Barthes.[71] LaCapra was the most significant teacher and mentor in my entire biography. For years afterward, I contemplated attending graduate school in history at Cornell and doing my Ph.D. with LaCapra. Ultimately, I rejected that idea and did not do it. Despite my affinity for LaCapra's ideas and passion for the subject matter, it would not have worked out well. I would have had an ambivalent "Oedipal" relationship with him as a father figure. It would have ended with my dramatically seeking to overturn the "authority of the father."

Modern European intellectual history, for LaCapra, was centered primarily in France and Germany. Secondarily, it was about developments in Great Britain, Italy, and Russia. Rather than divide landmark thinkers into separate humanities disciplines, he endeavored to see unity or commonality among literature, philosophy, psychology, social and political theory, and historiography. One gap in his curriculum was that natural science was left out. We did not even read Charles Darwin.[72] The lectures and readings in the nineteenth-century course began with the reflections on the French Revolution of Alexis de Tocqueville, Joseph de Maistre, and Edmund Burke.[73] This was followed by the German idealist philosophy of G.W.F. Hegel; the social theory of Karl Marx; the novels of Stendhal, Dostoevsky, and Flaubert; the positivism of Auguste Comte; the classical liberalism of J.S. Mill; the founding works of French sociology of Émile Durkheim; and the *fin-de-siècle* "God is dead" philosophy of Friedrich Nietzsche.[74]

In a "deconstructionist" way combining theories of history and literature, LaCapra taught that the author – no matter how "seminal," or perhaps because he was "seminal" – was not the sovereign master of his discourse. Two or more "currents" of thinking within the texts of a given writer were active and could be identified. These strands often co-existed in an interesting creative or "critical tension" with each other. Regarding Karl Marx, for example, LaCapra saw three different Marx-es. There was the quasi-religious faith in a teleological or eschatological narrative of progress in history in the "young Hegelian" Marx. There was the quasi-positivist economic determinism of the late Marx of *Das Kapital*.[75] The most interesting Marx was the open-ended critical theory model of the relationship between man, nature, and technology. Man's "species-being" is realized either in a salutary way in his "objectification" via activity that changes something in the world or degraded into "alienation" or "estrangement" under the capitalist conditions and relations of production. The critical theory model was most succinctly outlined in the *Economic-Philosophic Manuscripts of 1844*.[76] It inspired the Frankfurt School of neo-Marxist critical sociology of Theodor W. Adorno, Herbert Marcuse, and Walter Benjamin.[77]

LaCapra was fascinated by Flaubert's ambition to write "a novel about nothing." It was eye-opening to learn that the ascent to the apex of realism in the novel was undertaken in an ironic mode. Flaubert's artistic project was to achieve flawlessness in form. In a perfectionist novel like *Sentimental Education* and his satirical inventory of clichés and unreflectingly repeated ideas in culture, *Dictionary of Received Ideas*, Flaubert makes fun of the grandiose optimistic visions prevalent in Europe, as well as encyclopedic projects of classifying, listing, and recording all scientific knowledge.[78] Lionel Trilling writes in his preface to the English edition of *Bouvard and Pécuchet* that Flaubert's novel demonstrates that "the whole vast superstructure of human thought and creation is alien from the human person."[79]

The lectures and readings in the twentieth-century European Intellectual History course began with the psychoanalysis of Sigmund Freud and the founding works of German sociology of Max Weber.[80] This was followed by the language philosophy of Ludwig Wittgenstein; the novels of Thomas Mann, Virginia Woolf, Jean-Paul Sartre, and Albert Camus; the structuralist anthropology of Claude Lévi-Strauss; the Marx-Freud synthesis of Herbert Marcuse's *Eros and Civilization*; the history of reason and madness in Europe of Michel Foucault; and the critical theory analysis of science and technology of the early Jürgen Habermas.[81] Neglecting natural science, little to no mention was made of Albert Einstein's relativity theories or Niels Bohr's quantum physics.[82] In both semesters, there was an under-representation of women thinkers.

## Hiding Out in the Library Stacks

I spent much of my two years at Cornell in the stacks of the John M. Olin Library. It was a quiet place, isolated from everything worldly. It was the core experience of the proverbial Ivory Tower. I read or perused every book by (or on) Marx, Nietzsche, Dostoevsky, Sartre, and Camus that had ever been published in English.⁸³ I could not yet read any foreign languages. I often stayed up there in the stacks until nearly midnight.

My first romantic relationship with a woman that lasted for a significant length of time started during my first semester at Cornell when I was eighteen. It was with Karen, a talented poet and English major. She was with me in LaCapra's lectures and other literature courses. Karen, like me, was a lost soul. She had no idea what she was going to do with her life. We were together as a couple for one year. During the Spring semester and the following summer, we lived in a small room in a communal house in "College Town." After eight months, we continued our love relationship but went to separate rooms in different houses. I broke up with Karen after she betrayed me by sleeping with another man. I felt very hurt by that; perhaps a psychic injury semi-willfully intensified because it served as a convenient excuse for self-pity. Karen and I stayed friends for about ten years before losing contact.

Age 19, Ithaca, NY
(Alan N. Shapiro private collection)

Karen and I had three cats together. Two were black as the night, and the third was a small tiger cat. One after the other, they were hit by cars while wandering and died.

I went with Karen to meet her parents, who lived in a big house in northern New Jersey. Her father was an IBM computer salesman. At dinner, he gave a long speech about how much he loved the books of Ayn Rand and how she was the preeminent thinker of the twentieth century who had the most correct political and economic philosophy.[84] The following morning, I did something incredibly childish and immature. I placed one of his Ayn Rand paperback books in a small trash basket for discarded papers in his study. Karen's parents were initially very offended. But then I apologized. They forgave me, and all was well.

## Rupert Roopnaraine and Marxist Literary Theory

A second professor at Cornell who influenced me a lot was Rupert Roopnaraine.[85] Rupert was from Guyana and had recently earned his Ph.D. in comparative literature at Cornell with a dissertation on Charles Dickens' *The Pickwick Papers*.[86] LaCapra and Roopnaraine taught a graduate seminar on Flaubert and Sartre, having the students read Sartre's three-volume study of Flaubert called *The Idiot of the Family*.[87] Roopnaraine was a passionate Marxist revolutionary with a strong anti-capitalist and anti-imperialist worldview. Not only was Rupert brilliant, but what made him more interesting to me than typical orthodox Marxist intellectuals was that his main scholarly interest in Marxism, unusually, was in Marxist literary theory.

Rupert had a fantastic book collection, which I spent many hours poring over in his apartment. He had us read Mikhail Bakhtin's *Problems of Dostoevsky's Poetics* and *Rabelais and His World*, V.N. Voloshinov's *Marxism and the Philosophy of Language*, Lukács' *History and Class Consciousness*, and Lucien Goldmann's studies of Pascal and Racine.[88]

Rupert gave great lectures in a course called "The Modern European Novel." We read *Journey to the End of the Night* by Louis-Ferdinand Céline, *The Counterfeiters* by André Gide, and *The Confessions of Zeno* by Italo Svevo.[89] These novels deeply affected me with their literary qualities and the long and winding adventures of their antiheroes. Rupert wanted us to do close readings of Karl Marx's works, interpreting them as "writerly" texts of great literature.

Rupert gave up his promising academic career in comparative literature to return to Guyana as a political activist. His friend and countryman Walter Rodney, author of the acclaimed post-colonial historical work *How Europe Underdeveloped Africa*, was also at Cornell while I was there.[90] Rodney was assassinated in a car explosion in Georgetown in 1980. Rupert became the leader of the Working People's Alliance political party. From 2015 to 2017, he was the Minister of Education of Guyana. Rupert is also a leading scholar and advocate of the literature of the Caribbean. After leaving Cornell, Rupert taught literature courses at Columbia University. As a member of a South American political party the U.S. government classified as "communist," Rupert was no longer allowed to enter the United States.

Another professor who influenced me was John Weiss, who taught Twentieth-Century European Social History.[91] There were great discussions in Weiss's seminars. He had us read several books explaining the rise of Naziism in Germany in the 1930s. We also read Hannah Arendt's *Eichmann in Jerusalem* and the novels *All Quiet on the Western Front* by Erich Maria Remarque and *Christ Stopped at Eboli* by Carlo Levi.[92] I understood how the literary novel can profoundly witness political and social history. I took other seminars at Cornell Arts and Sciences in philosophy, government, and American, French, and German literature.

## The Leftist "Student Movement" at Cornell

With my close friend and roommate Marc, I was in a seminar in the government department (the equivalent of political science) on "Latin American Politics." The professor was the noted scholar Eldon "Bud" Kenworthy, a great guy.[93] Kenworthy died in 1998 in Walla Walla, Washington, after he was hit by a car. There were about fifteen students in the seminar. Most would pursue financially lucrative careers in business, law, university administration, or public policy. But at this stage in their lives, all of them (except me) identified as radical left Marxist revolutionaries. A major topic for discussion in the group was the recent 1973 military coup d'état in Chile – notoriously supported by U.S. President Nixon and soon-to-be Secretary of State Henry Kissinger – which violently overthrew the democratically elected government of the democratic socialist Salvador Allende. After reading journalistic books and articles about the circumstances and events in Chile, most of the students concluded that Allende had "pussy-footed" too much in his dealings with capitalist powers and economic interests. It would have been better to forcefully take control of the situation with more violent measures, as advocated by the "urban guerrilla warfare" Revolutionary Left Movement (MIR).

The students in the seminar idolized the martyred Marxist revolutionary Ernesto "Che" Guevara. After reading the book *My Friend Che* by Ricardo Rojo, we were assigned to write ten-page papers that would be available in the library for everyone in the group to read.[94] I wrote my paper on Orwell's and Camus' critiques of top-down state socialism from their libertarian socialist perspective.

To me, the Cuban Revolution was misguided state socialism. We spent one entire two-hour session of the seminar discussing my paper. It was a fierce battle of one (me) against fifteen. They "ganged up" on me. My friend Marc took my side or defended me to a certain degree, staking a neutral position in the debate.

The Cornell campus had a social "scene" of leftist political activism. This "student movement" consisted of about two hundred people. There were Marxists and anarchists; ecologists influenced by the "post-scarcity" ideas of Murray Bookchin or the "small is beautiful" principle of the economist E.F. Schumacher; Trotskyists and Maoists; adherents of the Democratic Socialists of America or the Young People's Socialist League, and members of the Afro-American Society.[95] There were independents like me who mainly just wanted to read books. Twice, we occupied Day Hall, the university's administration building, and the location of the Office of the President. The first time, the occupation was ostensibly to protest an undergraduate tuition increase. I think we just wanted to express ourselves. Marc became a spokesperson of the "scene" and was interviewed on the radio about the intentions of the demonstrators. The second time, in April 1976, was about the conditions of financial aid for African American students and, more generally, the struggle for black civil rights. That occupation was even reported in the *New York Times*.

Another friend I made in Cornell's "leftist" student movement scene was the mysterious Mondovi. He had a wild, untamed look, with scouring pad-like Afro hair. His nickname was "Brillo." He was very sociable, and people constantly followed along after him in anticipation of something exciting happening. I was in a political science seminar together with Mondovi. We were assigned to work on an oral report together for class. One evening, we made an appointment for me to visit the house just off campus where he lived to research and prepare our presentation together.

It was almost winter. I followed a long, narrow cobblestone path through a swampy area, past some corporate buildings, then across the cemetery to get to College Town without wasting time taking the long route through the streets. It was raining hard all day. Even though I had an umbrella, my feet were getting soaked. I walked quickly but couldn't help but notice the tombstones around me, which made me feel a sense of the destiny of the hour at hand. I had only ventured into College Town a few times. This was certainly my first time being invited to a house there. The guys in the dorm thought of College Town as a den of iniquity where men and women lived together, no one went to classes, and there was fear of getting busted by the police for smoking pot.

Turning the corner where the cobblestone path rejoined the main road, I noticed the absence of people in the street. I went into a Copy Center to ask for directions to my destination address. A stocky employee with a fat black mustache, wearing an oversized blue work jacket, smiled sardonically. He muttered a few curt directions. I thanked him in a soft, low voice. I glanced at the magazines lined along the back wall. I went outside, no more confident of how to reach my journey's end than before. I started downhill, passing a blacktop parking lot full of potholes. I hit the right street almost by chance. There was a Shell station on the corner that I vaguely remembered Mondovi mentioning the day before.

An amiable-looking fellow with long red hair dressed in jeans and a bleached T-shirt marred by a big green spot came to the front door to answer the bell and asked me what I was looking for. I was led down a dark stairwell into a small hallway with a big old oak desk with mailbox pigeonholes above it. It was a wing of the house set apart from everything else. The red-haired fellow knocked firmly on the door next to the seldom-used desk and yelled: "Hey, Brillo! You in there?" We were soon inside a large, square-shaped, sparsely furnished room. There wasn't much in it other than several multi-colored tapestries hanging from the ceiling and a few grey and black cats who darted in and out through an open window. Mondovi was seated on a navy blue beanbag in the middle of the floor, looking at the song texts on the back of a record album. Two women were slouched on either side of him, their knees deep in the acrylic carpeting that stretched from wall to wall, peering over Mondovi's shoulders. The music corresponding to the text emerged from all corners of the room. The notes and chords were unknown to me.

Mondovi rose from the bean bag and shook my hand. "How you doing, man? It's raining so hard I figured you'd call up and tell me you weren't coming. Take your shit off and warm up for a while." Mondovi was quite talkative. Soon, he launched into a detailed explanation of his living arrangements, the various characters in the house, the sexual relations that were going on, and the crazy thing that happened at the house party last Saturday night. Several people were so plastered that they thought nothing of it when, at 2 AM, a drunk driver swerved off the road in front of the house, his car ending up on the porch outside the living room. Assuming that it was just another late-arriving guest, the inebriated revelers helped the driver (who may have twisted his neck) out of his car and offered him some punch. He joyfully accepted and, within a quarter of an hour, was among those who passed out on the living room floor. When he "came to" the next morning, he gathered up his belongings and took off in his car, the only uncertainty being whether to go to the hospital or the auto repair shop.

Everyone laughed at this story, but Brillo insisted that it wasn't meant to be funny. It was damn serious and should make them all think about "where their heads were at." Brillo introduced me to the two women. One said I was cute, although I was small and looked very young. I could not hide my embarrassment and blushed despite trying not to. I didn't think that I looked so young anymore. My unease was apparent to everyone. Both women started to laugh again. Brillo got angry and told them off. He quickly changed the subject to break the silence that followed. I knew at that moment that I had found a friend. Everyone in the group talked about what we were studying and would do after leaving Cornell and Ithaca.

The red-haired fellow, who had remained in the room, rolled up a joint and began to pass it around. In the dimly lit room with the music's distraction, I hoped nobody would notice when I put the cigarette to my lips without breathing in. I fidgeted nervously as it came closer to my turn. The woman who deemed me cute handed the reefer to me. Just as I raised it to my mouth, she declared, "You've got to inhale, baby!" I snorted non-consciously. A huge quantity of smoke rushed down my throat and burned my lungs. I coughed violently.

The women started chuckling again. Brillo (Mondovi), who had gone quickly to the bathroom to get a glass of water, looked at them sternly. I tried in vain to regain my composure. I looked down at the floor. He grabbed my hand. Soon, we were outside in the drizzling rain, walking deliberately. I wondered about Mondovi's true reasons for befriending me, but the thought passed quickly. We entered an almost empty bar. There was no telling where everybody was on this rainy November night.

> Brillo motioned with his chin for me to sit at a corner table while he went to get drinks. When he returned, I was stroking my forehead with my bent knuckles, one elbow on the table, lost in contemplation. Neither of us said a word since we left the house. "I usually don't tell people this, but when I was a kid, I skipped two years in school." I'm only eighteen. These things are new to me." I spoke falteringly, seeing what Brillo's reaction would be.
>
> After a few seconds of silence, he replied: "You know, the type of life we have, middle-class American suburban life, is not usually the stuff of which literature is made." I sat back in my chair with a frothing glass of beer and recounted my childhood.

## Looking Towards France

In 1952, the friendship between the two French luminary existentialist thinkers Jean-Paul Sartre and Albert Camus ended after Sartre disagreed vehemently with the critique of Marxism and Soviet communism that Camus articulated in his book *The Rebel*. In the late 1950s, Sartre and Camus also had diverging positions about the Algerian War and the anti-colonial independence movement. In evaluating these important debates in twentieth-century French intellectual history, I am on the side of Camus in his rejection of the terrors of U.S.S.R. state socialism and the violence of the Algerian National Liberation Front, which resulted in an estimated one million deaths. I do not like Sartre's fellow-traveling and pretzel-logic defense of Stalinism.[96] Nor did I care much for the Maoism that Sartre embraced in the aftermath of the Paris student uprising of May-June 1968. However, there was one aspect of Sartre's system of thinking which was very important to me and which – for better or worse – had a profound effect on the decisions that I made in my life. This was Sartre's concept of the *engagement* or "commitment" of the radical intellectual.

Just as man or woman is condemned to be free, to make open-ended choices without knowing absolute right or wrong, and without any signs or verification from the non-existent God, one is also condemned to be "engaged." We are "engaged" whether we like it or not. It is a given "fact" that is endemic to the human condition. It is the moral duty of the thinker to take part in history. As a leftist thinker, Sartre faced the challenging paradox that he sympathized with the plight of the working class and the subaltern yet was himself of "bourgeois" origin. Self-questioning of his status in the capitalist system was essential for authenticity.

Sartre refused the Nobel Prize in Literature when he was awarded it in 1964. I considered this an exemplary gesture symbolizing the refusal of privilege and celebrity. The existentialist Marxist ethics of Sartre seemed to me to imply the rejection of the favors of elite institutions. Suppose you believe that we live in an unjust class society where the rich, privileged elites exploit and oppress the poor and the working class. In that case, you should not accept the advantages offered to you by elite institutions – like universities! I don't know if American Marxist intellectuals who are so concerned about inequality and alienation yet accept the elite privileges of their professorships at Ivy League and similar universities are hypocrites. I am reluctant to judge that. I am not going to make a general ethical statement about it. And to do so is not at all my point. My point is about what this reflection meant for me in my auto-socio-biography. It was a seeming hypocrisy that I could not accept myself. It felt, and has always felt, like a deceitful path to take with my life. That is why, for example, I turned down a full fellowship I was offered to do my Ph.D. in the politics department at Princeton University.

Given this hesitation to accept the future charted and available to me in line with my academic excellence, I was a lost soul facing American society at the age of twenty. On one side, alienated labor was too anathema to me to want to do any full-time, forty-hour-a-week job. On the other side, I did not want to partake of the entitlements of an academic career. I would have to find some "third" arrangement or setting to make my way. Additionally, it became clear in America in the late 1970s and early 1980s that the only way to land a professorship in the humanities or social sciences would be to do one's Ph.D. at an Ivy League-level university. To do graduate studies at a mid-level university would get one nowhere. I was constantly told that it was not even worth trying. All the teaching/research positions were taken by the early postwar Baby Boomers and participants in the 1960s counterculture who were ten years older than me.

I fantasized about becoming a proletarian, living authentically without bourgeois birthright, to survive financially, perhaps with a part-time job. It is to be something like what the Italian Marxist thinker Antonio Gramsci called an "organic intellectual": living the ordinary life of the workers while contemplating and writing about it. I wanted to live in "civil society" without being a professor or having a full-time career, yet earn enough money, probably through "menial" jobs, and do my research and writing at night in a small study in the attic.[97] It was a crazy idea, yet I pursued it over the next decade. It was impossible because, in an age of permanent economic inflation with everything – apartment rents – always becoming more expensive, how could you keep the monkey off your back?

The American radical thinkers were all Marxists. In Spring 1976, my last semester at Cornell before completing my Bachelor of Arts degree, I discovered French thinker Jean Baudrillard. I read Baudrillard's book *The Mirror of Production*, which had just been published in English translation by Telos Press.[98] I grasped that Marx was not radical enough. Marx unwittingly mirrors the logic of capitalist industrial production, which he intends to criticize. I realized that the sort of "religion" of production or economic growth that Marx makes goes against the grain of sensibility to the ecological destruction of the planet and the "limits to growth," which was one of my concerns and values. In the final section of *The Mirror of Production*, entitled "The Radicality of Utopia," Baudrillard exhibits the anarchist dimension of his thought – and comments on the student uprising of May-June 1968 in France.

Six months later, after I finally learned enough French to read books in that language, I read a second book by Baudrillard: *The Consumer Society: Myths and Structures*.[99] He writes about consumerism, shopping mall architecture, television, and advertising. For Baudrillard, the mass identity architecture of hypermarkets, department stores, cookie-cutter houses and apartment complexes, international airports, and hotel rooms is the "deterritorialization" of the "terrain" of physical space. Baudrillard zeroes in on the fetishism of the consumable object and the "false" abundance of consumerism. The alleged abundance of the society and economy of growth is the mirage of the spectacle of the accumulation of commodities. The critique of the consumer society begins with the critique of the supposed "naturalness" of "needs" and their satisfaction in all human economic systems, which is the "universal model" of "bourgeois" economic theory and its justifications of capitalism.

Jean Baudrillard (Photo by Ayaleila, Wikipedia Creative Commons BY-SA 4.0 License), Guy Debord (Public Domain)

At an indeterminate point in history, America exploded from physicality to virtuality and "became the world." Or the other way around – the whole world imploded into virtuality and became Americanized: the same big color TV screens everywhere; the same shopping malls; the same Coca-Cola, all-American hamburger, and "French fries"; your identity, logo, or "personalized" message printed on a T-shirt; horseback-riding cowboys and Superman comics; Elvis Presley and Marilyn Monroe; eventually McDonalds, Starbucks, Nike sneakers, and Apple computers everywhere on the planet.

The social theory which interested me was a critique of everyday life in what, in the anarcho-Marxist literature, was called the "advanced capitalist society." The critique was not only about America – but about all the "first world" countries. The rebellious and critical spirit of the movements of 1968 died out in America. I believed that I might still be able to find it in Europe, and specifically in France. I started to make plans to spend time in France. I discovered the artist-activist group called the Situationists and their amazing book *The Society of the Spectacle* by Guy Debord.[100] I might meet Situationists in San Francisco or Paris.

Debord was a principal figure of the Situationists, a movement of radical artists, architects, writers, and political activists prominent in Paris and other French cities like Strasbourg, Amsterdam, London, and many towns in Italy from the 1950s to the 1970s. Debord was a neo-Marxist seeking to comprehend how capitalists' control over workers' lives expanded from the sphere of production to consumerism, organized leisure, daily life, and the media culture of images and rhetorical language in late capitalism.

With his concept of *the spectacle*, Debord understood that the omnipresence of images institutes a world of abstraction and acquiescence, diminishing what is "directly lived" and an increase in the autonomy and power of the screen and the images themselves. Something becomes true – or *more true than true* – by having been said charismatically in the media. In the *spectacle*, "the liar has lied to himself." "In a topsy-turvy world," writes Debord, "*the true is a moment of the false*."[101] Social life shifts from *being* to *having* to *appearing* – ending as the reign of appearances. Debord writes:

> In societies where modern conditions of production prevail, life is presented as an immense accumulation of spectacles. Everything that was directly lived has receded into a representation.
> 
> The images detached from every aspect of life merge into a common stream in which the unity of that life can no longer be recovered. Fragmented views of reality regroup themselves into a new unity as a separate pseudo-world that can only be looked at. The specialization of images of the world has culminated in a world of autonomized images where even the deceivers are deceived. The spectacle is a concrete inversion of life [...]
> 
> The spectacle cannot be understood as a mere visual excess produced by mass-media technologies. It is a worldview that has actually materialized and become an objective reality [...]
> 
> [The spectacle] is both the result and the project of the present mode of production. It is not a mere supplement or decoration added to the [alleged] "real world." It is the heart of this real society's unreality. In all of its particular manifestations – news, propaganda, advertising, entertainment – the spectacle is the model of the prevailing way of life [...]
> 
> In both form and content, the spectacle serves as a total justification of the conditions and goals of the existing system.[102]

Situationist ideas and practices were prevalent in France in May-June 1968 and in the San Francisco-Oakland-Berkeley Bay Area in the United States. Situationism produced works of radical utopian architecture like Constant Nieuwenhuys' New Babylon project, a science- fictional worldwide city of the future.[103] The Situationists elaborated the idea of "unitary urbanism," the dream and design of a city of endlessly enchanting and participatory non-functionalist *situations*, among which creative-passionate citizens would experientially drift.

As described in Chapter One, I booked my plane flight to Luxembourg for July 3, 1976, one day before the ultimate patriotic American "Society of the Spectacle" bicentennial celebration. I had been to Montreal three times, to Puerto Rico and Aruba for winter vacations with my "bourgeois" Long Island parents, and to

CHAPTER 6

Tijuana, Mexico, for an afternoon. But, other than those excursions, I had never left the country. Before crossing the Atlantic Ocean, I decided to make a quick four-week trip to California, where I had only been once, at age seven.

When I was eleven, in 1967, my parents took my brother and me to Montreal for one week to the Expo 67 World's Fair. We were there on the opening day of the International and Universal Exposition, April 28th. I photographed the Russian ambassador to Canada cutting the ribbon to inaugurate the Soviet Union pavilion. I took a picture of the French pavilion.

Opening the U.S.S.R. Pavilion, the French Pavilion, Montreal Expo 67 (Photos by Alan N. Shapiro).

I had little money, so I bought a Greyhound Bus ticket from Ithaca, New York, to San Francisco, California. The exhausting trip was going to take forty-eight hours. Somewhere in Ohio, a low-life, grungy, tough-looking man about five years older than me sat down next to me. He told me that he had just gotten out of prison. He had a great time there because he did several young men weaker than himself in the ass. I was horrified. He said he would do it to me in the men's room at the next highway restaurant rest stop. I was terrified. I stayed on the bus during the rest stop. This was unpleasant because I had to endure continuous hours before and after that incident without getting off the bus. I was hungry, but that became a low priority. I changed seats to get as far away as possible. When the ex-convict returned to the bus, he looked around but, fortunately, did not see me.

The bus trip was stressful, so I interrupted the straight shot to the West Coast. I took a bus from Minneapolis to Kansas City, Missouri, and did a two-night and one-day visit with my uncle (my father's brother) and his family. I got off the Greyhound bus in Reno, Nevada, and ventured into a gambling casino for the first time. I only looked around. I arrived at the San Francisco bus station and found my way to nearby Berkeley, where an old friend from MIT named Dan was now a law student. I had been in political science classes with Dan, and we had stayed in contact. He let me sleep on his living room sofa. His roommate was a former tennis counselor at the camp in New Hampshire where I had spent four summers.

I was familiar with the Berkeley campus from the film *The Graduate*. I attended a few lectures on European Intellectual History with Professor Martin Jay, the author of the acclaimed book about the Frankfurt School and the Institute for Social Research called *The Dialectical Imagination*.[104] Jay's approach to the history of ideas was not as deep as that of LaCapra. Cody's bookstore on Telegraph Avenue in Berkeley had a large section of anarchist and Situationist books and pamphlets. I had never seen that in any bookshop before.

I went to two meetings of a Situationist collective and reading group in San Francisco. I felt a sense of belonging and camaraderie there. Their ideas were closer to mine than anyone in the Cornell leftist community. But still, I felt that joining any organization would be a bad idea. I avoided, like the plague, anything that resembled a "cult" or a "sect." I went to jazz clubs in San Francisco in the evening. I ate many dinners in Chinatown.

I spent a week in Los Angeles and Santa Monica. I slept on the sofa in the apartment of my friend Mark, my housemate from Cornell. Mark was a dear friend, a philosophy major dedicated to Wittgenstein and dense texts of Eastern religions. We went for long walks and had sincere personal discussions. I found a ride in a car with three others from Los Angeles to St. Louis from a ridesharing bulletin board. We stopped at the Grand Canyon along the way. I hung out in the library of Washington University. I found another ride from St. Louis to New York City. I borrowed a car from my parents, drove to Ithaca one last time, and got all my stuff and books. I drove back to the house in Roslyn. I went to a bank and bought my twelve hundred dollars in traveler's cheques. I had a few last decent home-cooked meals from my mother. Then the day arrived. It was time to take wing to Europe. My Dad drove me to the airport.

"I was twenty. I will let no one say it is the best time of life."

CHAPTER 7

# First Year in France and Italy

What was unusual, and in a way crazy, about my plan was that I was going to France to simply "live" there. I had no job. I had no affiliation with any university program. At least I had the project of learning French. That would give me something concrete to devote my time and effort. My trans-Atlantic flight stopped in Reykjavík, Iceland. I exited the plane, walked around the airport, and ventured outside for a few minutes to breathe the far north air. It was much cooler than the heat wave I left behind in New York.

Sitting next to me on the plane was Juan Pablo Rivera from the South American nation of Colombia. He changed planes at JFK Airport. Juan Pablo was a couple of years older than me. He did not speak any English. Despite getting A's in Spanish in high school for two years, I could not speak Spanish. Neither of us could say more than a few words in French. I quickly realized that Juan Pablo was my literary alter ego. As far as I could tell, he was going to France for similar reasons. He just graduated from his university in Bogotá. He was disillusioned with his country. He saw no future for himself there. He was passionately interested in French literature and philosophy. He wanted to connect with the student movement's post-1968 culture of *le gauchisme* [leftism]. He wanted to learn French. After the plane landed, we said goodbye. But a few weeks later, I would, by chance, again encounter Juan Pablo.

My immediate destination was Madrid, Spain. My ex-girlfriend Karen, with whom I was still on friendly terms, had just done a "semester abroad" there. I would meet with her in the Spanish capital and have her company for a week. My plan after that was to cross the border back into France alone. I did not know a single person in France. I had the address of a second cousin from New York named Jeffrey, an expatriate living in Poitiers. I had never met Jeffrey but was considering contacting him at some point.

## CHAPTER 7

I made it from the Luxembourg Airport to that city's main train station. Then I took a train to the Gare du Nord in Paris. From there, I took the Métro to the Gare de l'Est, where trains departed to Spain. Unfortunately, it was late at night, and the next train was at about 6 AM. From midnight on, I sat quietly in a waiting room. I was afraid to fall asleep, fearful of possible theft of my small backpack or sleeping bag. I played chess for several hours with a male French university student. We could barely speak each other's language.

The trip south through France the next day was exhilarating. It was an old rusting train. Much of the time, I was alone in a six-seat compartment. I had the window wide open. It was hot summer weather. The countryside was beautiful. There were fields and farming tracts as far as I could see – the loud noise of the train with the open window added to the palpability of the experience. I recklessly stuck my head out the window to feel the refreshing breeze. Only later would I take seriously the ubiquitous printed warnings to *Ne pas se pencher au dehors* (Don't stick your head out the window). I am in Europe! I am in France! I am no longer in America! I shouted it out the window. It was the greatest feeling in the world.

We arrived in Irun, just over the France-Spain border. From there, I would get a train to Madrid. Spain's longtime fascist dictator, Francisco Franco, died in November 1975. The country was at the very beginning of its transition to democracy under the constitutional monarchy of King Juan Carlos I. At the Irun train station, there was a crowd of hundreds of travelers, many of them young people and hippies, milling about in semi-panic and disorientation. There was a long line to get to the three open ticket windows. It appeared that one would have to wait hours to reach a sales counter. Several people told me that my ticket from Luxembourg to Madrid was only valid for the slowest train, which would take ten hours to get there, and only departed a few times per day. One needed a supplementary ticket to get on a faster train that departed more often and would reach Madrid in about five hours. Moreover, it was not possible to pay the supplement on the train. There was a fast train leaving in about thirty minutes. I was standing with about ten English-speaking young people (Americans, British, Australians) on their way to Madrid. The group decided to get on that next train and see what would happen. It was late afternoon. No one had the patience to wait in the long queue to buy the supplementary Madrid ticket.

I took a seat next to a window in a second-class car. Soon, a conductor came by. He looked at my ticket, shook his head angrily, and shouted at me in Spanish. He motioned for me to stand up and grab my stuff. He led me to the narrow space between two cars. It was clear that he would kick me off the train at the next local stop in the middle of nowhere. He returned moments later with a second supplement-less truant fish he caught. Then, a third and a fourth. He stood guard patiently over us. We assumed he didn't understand English, especially if we talked fast, so we plotted among ourselves. When the train stopped at the next backwater *pueblucho* [small town], we would get out, run as fast as we could two or three cars back, then hop on again. We would disperse and try to become invisible. The sense of adventure in what we were doing was as much of a motivation as the practical reasons.

I retook a seat. No conductor came by until about twenty minutes before Madrid. I almost made it without further trouble. But not quite. I couldn't escape another check of my ticket, and soon, a different conductor scolded me loudly in Spanish. Now, he was willing to let me pay for the supplement on the train, which did not seem possible before. But the next problem was that I did not have any Spanish pesetas. I planned to buy some as soon as I got to Spain. But, on top of everything else, it was Sunday, and the currency exchange office at the Irun train station was closed. I offered the angry conductor U.S. dollars or French francs, which only made him angrier. Finally, a generous Spanish man witnessing the scene resolved the situation by paying for my ticket supplement. I profusely thanked him and offered him U.S. dollars as reimbursement, but he refused to take my money.

The train finally arrived in Madrid. I had Karen's phone number. I found a pay phone and called her. She came to the station to pick me up. She arranged a room for me in a cheap hotel. Karen knew inexpensive places to eat. The Spanish food was tasty. Spending a week with Karen before setting off on my own was comforting. Then the time was up.

I took a train early one morning to the French coastal city of Bayonne. Bayonne had some meaning for me because the literary theorist Roland Barthes spent part of his childhood there. I walked around Bayonne for a few hours. The town of bright summer colors next to the water was enchanting. It occurred to me that it was again Sunday. I saw French families out for their traditional day-of-rest strolls. Contrary to them, I was now facing an indefinite period of time alone. It was a nervous feeling.

I got on another train to Toulouse, the largest city in southwestern France. It was late evening when I arrived. I shared a hotel room with bunk beds the first night with some young New Zealanders I met on the train. On Monday, I wandered around the city center, searching for an affordable room for a few weeks. A man at a hotel reception on the *Rue du Poids de l'Huile* facing the *Square Charles de Gaulle* showed me a tiny maid's room up some rickety stairs above the hotel's upper floor. There was a single bed and a small desk, nothing else. The bathroom was downstairs. He offered me a cheap weekly price. I moved in there with my few possessions. It was time for my self-organized immersion course in French to begin.

## Summer in Toulouse

For six weeks, I spoke nearly no English. My Bible was the *French Grammar* book by Francis M. Du Mont from the "Barnes & Noble Outline" series.[105] I drilled verb conjugations into my head for several hours each day. I went around the city listening to snippets of other people's conversations, for example, in museums and on park benches. I bought baguettes, cheese (*la vache qui rit*) [the laughing cow], yogurt, and small packets of *le jambon* [ham] and ate in my room. I was lonely.

One afternoon, I was walking in the public square across from the hotel, and there was Juan Pablo Rivera, who was next to me on the Icelandic Air flight, sitting on a bench. We were thrilled to see each other and flabbergasted by the coincidence. Perhaps he mentioned during the flight that he was going to Toulouse, so I unconsciously chose that city for my summer language "internship." Juan Pablo's French had improved, and so had mine. We were able to communicate better. An English native speaker and a Spanish native speaker weirdly speaking with each other in broken French. Our friendship was formed and deepened. After that chance meeting, we met up every second or third day. Juan Pablo was staying with a French couple whose address he had from mutual friends in Bogotá.

I was alone and found some comfort in looking at the front and back covers of French books in the many bookshops of Toulouse. I could now see the tomes in the original language of the novelists, philosophers, and political thinkers whose works I was previously familiar with in English translation. There were many eye-catching series of intellectually rich yet monetarily inexpensive paperback books. While glancing at a book in the anti-colonialist Marxist Éditions François Maspero series, I conversed with a young Frenchman, Pierre, who was one or two years younger than me.[106] We went for a coffee and chatted for two hours. He spoke no English, so I stumbled along in French. Pierre was well-dressed, wore glasses, and smoked a pipe. He just finished his final year at the *lycée* and was slated to go to Paris in September to study at the École Normale Supérieure. He was extremely well-read. We discussed many authors and the evils of capitalism. Pierre was from a wealthy family. His father was a top manager at Aérospatiale Toulouse, part of the state-owned company involved in the French-British joint project of building the Concorde supersonic airliner. Pierre was a bit snobbish but took a liking to me. He felt bad for how little money I had. During the six weeks I spent in Toulouse that summer, we met twice a week, and he bought me lunch in restaurants.

I had separate occasional meetings with Juan-Pablo and Pierre, my essential yet tedious French grammar studies, whichever book in French I was trying to read then, and my long walks around Toulouse. Yet I was still lonely and a lost soul. One afternoon, I was at the hotel, climbing to my *chambre de bonne*, when I heard music from Pink Floyd coming from one of the rooms. I never realized that young people in Europe listened to Pink Floyd. It sounded to my forlorn ears like salvation or homecoming.

*Us and Them* (1973), by Pink Floyd

Us us us us us and them them them them them
And, after all, we're only ordinary men
Me me me me me and you you you you you
God only knows, it's not what we would choose to do

Haven't you heard, it's a battle of words
And most of them are lies
Listen, son, said the man with the gun
There's room for you inside (Harvest label)

*Brain Damage* (1973), by Pink Floyd

And if the cloud bursts, thunder in your ear
You shout, and no one seems to hear
And if the band you're in starts playing different tunes
I'll see you on the dark side of the moon (Harvest label)

I figured whoever was in that room must be a fellow person from the English-speaking world. With the excuse of the music, I gathered my courage and knocked on the door. It was a friendly young French hippie with long blonde hair and a beard. After brief introductions, he told me that he had an American friend in his evangelical religious community whom I could meet. The congregation was meeting that evening, and I could go with him if I wanted.

It sounded strange to my atheist mind, but I was lonely enough to agree to go. That's how I met Katherine, a woman from Minnesota who had been in Toulouse for three years on her evangelical mission. Over the next few weeks, I met many times with Katherine. I attended the religious services of her congregation. I listened to the sermons while thinking to myself, what nonsense. Yet it was a good way to listen to a lot of French. They all sang together:

Nous disons merci, oh merci Seigneur!
Nous te disons merci pour toutes tes merveilles
Tu les as inventé pour nous, pour la beauté des choses, et la bonté des êtres

# CHAPTER 7

We say thank you, oh thank you, Lord!
We say thank you for all your wonders
You invented them for us, for the beauty of things, and the goodness of beings

Tu nous as donné la Terre à garder et cultiver
Merci pour les nuits paisibles et les cieux étoilés
Merci pour le rugissement des lions et le chant des oiseaux
Merci pour la douceur des fruits and le sel des océans
Nous disons merci, oh merci Seigneur!

You gave us the Earth to keep and cultivate
Thank you for the peaceful nights and the starry skies
Thank you for the roar of the lions and the song of the birds
Thank you for the sweetness of fruits and the salt of the oceans
We say thank you, oh thank you, Lord!

Katherine had an idea of how I could save money. I could move for the whole month of August into a studio available in an old apartment building. I would live across the hall from Monsieur Pierre Peota, a frequent attendee at the religious services. After settling in, I spent a lot of time with Peota. He took me on bus rides to the countryside, showed me how to pick wild mushrooms safely, and introduced me to French suburban hypermarkets where you could get a coffee and a *croque monsieur* at the cafeteria. M. Peota was fifty-five years old. He looked older but was still physically fit. He had been a postal worker. One time, he came home in the middle of the day and found his wife in bed with another man. He grabbed a sharp kitchen knife and stabbed both his wife and her lover many times. Fortunately, no one died. M. Peota was proud of what he had done since, in France, it was considered a justified and even honorable crime of passion. He was sentenced to four years in prison.

Pierre P. confided in me that he was an atheist. He came to understand that he could get released from prison early by faking an emotionally charged religious conversion and embracing Jesus Christ as his savior. He recounted in detail how he staged his performance of declaring himself a sinner and announcing his reborn salvation. But the price to be paid for having been sprung from the slammer on religious grounds was that he had to authenticate his embrace of God by becoming an active member of a Christian community.

I wrote a letter to my second cousin, the American expatriate in Poitiers, and he replied. Jeffrey was a translator of high French literature and survived with very little money. He explained in his letter that every year in September, he made some cash by picking apples at a farm (like the seasonal *vendanges* where you pick grapes). Juan-Pablo and I could accompany him and try our hand at transferring pomaceous fruit from trees to large crates.

The rent I paid covered my stay in the studio apartment in Toulouse until August 30th. Juan Pablo and I set off for Poitiers by train to meet up with Jeff on the morning of September 1st. On the evening before our departure, I was invited to dinner by an educated middle-class couple who lived in the flat next to mine. I conversed fluently in French with them for several hours on many different subjects. At the end of that evening, I said proudly to myself: hey, you now know French! I had gone from zero language knowledge to fluency in six weeks. Later in life, it would take me ten years to achieve a similar competency in German.

## A Quick Exit from France

I was a terrible apple picker. You had to remove the apple from its branch with a specific hand motion to break the stem from the bush while keeping it intact. I had trouble mastering that technique. An apple without its stem was worthless and had to be discarded. You were paid according to the number of giant crates you filled. The best pickers filled five crates per day and made some decent money. I worked my ass off and could only manage to fill two crates a day. It was hard physical labor, and I could not keep up. About thirty temp apple pickers slept on low single beds in a barn. They fed us three meals daily. The lunches and dinners were very good. There was rabbit stew, horse meat, and mussels in white wine sauce. But you had to wake up at 6 AM every morning. Breakfast was bread and coffee. The work was five and a half days a week. The routine was too grueling for me. I quit after two weeks and left the farm.

I hitchhiked back to Poitiers and found a temporary room for very little money in a student dormitory at the university. While sitting in a lounge reading books, I met Sally, a British and Jewish student of French literature. She was highly intelligent and very pretty and sexy. We started a romantic relationship, which, twenty-four hours later, became erotically intimate. Sally was the second woman with whom I made love. The first time that we had intercourse, a lot of blood came from her vagina. I thought that meant that her hymen broke, and it was her first time. But she vehemently denied having been a virgin. She told me at the start that our relationship would only last a few weeks. There was a French male student with whom she was in love. He would show up in Poitiers in October, and she would be with him. She spent a few weeks with me to practice having sex so she would be experienced when he arrived. She would avoid having to tell him that she was a virgin.

Sally rented an apartment. I spent time with her until her more "permanent" boyfriend appeared. I lost contact with her and never found out if their relationship lasted.

## CHAPTER 7

I spent a few weeks in Paris. I slept on the sofa in the room of Pierre, the pipe-smoking literary intellectual from Toulouse, in an upscale dormitory of the École Normale Supérieure. We argued about authenticity and elitism. That ended our friendship. I spent several weeks in Nantes, staying in the apartment of some leftist students I met at the apple farm. I slept in my sleeping bag on the floor.

In late November, my best friend Marc, whom I knew in high school and who was my roommate and fellow humanities student of Professors LaCapra, Roopnaraine, Kenworthy, and Weiss at Cornell, arrived.[107] Our emotional reunion happened at the Nantes train station. Marc's reasons for coming to Europe were similar to my own. He wanted to learn French. He wanted to meet groups of students and young people with activist critiques of the "advanced capitalist society." He wanted to test the waters of emigrating to somewhere in Europe. After a half-year of being alone almost all the time, it was a relief to have Marc's camaraderie.

After in-depth discussions, we came to the shared view that the post-1968 leftist student and counter-cultural movement that we were keen to hook up with was dead in France. Yet, based on stories we heard, it was still alive and thriving in Italy. Moreover, I did not find France to be especially friendly or hospitable. It was difficult for me to make connections with people. I was glad that I learned the language and could now read books in French. That was my accomplishment. Besides that, France was a dud. I did not see any chance of being able to settle and live in that country. Marc and I formulated a new plan. We would give Italy a try.

Before extending our horizon to Italy, we both wanted to continue to work on our French. We traveled by train to Avignon in the southeast to stay in France and get closer to Italy. We met a very amicable British professor with an apartment to rent on the second floor of his house. He agreed to let us stay there for just one month. We used that apartment as our springboard for our first excursion into Italy. We approached Italy like the Apollo mission, first orbiting the moon before returning on the next mission and landing there.

Italian trains were cheap, but my original twelve hundred dollars had dwindled, so hitchhiking was preferable when possible. We made a brief journey from Avignon to Venice and Bologna. I had only the equivalent of a hundred dollars with me. I left the rest of my traveler's cheques in Avignon. Not having all my "vast" capital with me was my way of rationing funds. We stayed in Italy for a few more days than expected. At the end of the sub-trip, I had fifty centimes left in my pocket. Marc and I separated to maximize our chances of getting picked up by friendly car drivers. I bought my last baguette and, at the side of the highway, settled into the cold and darkness of the late December evening. My last hitchhiking ride got me to within twenty-six kilometers of Avignon. I had to walk the rest of the way. I was wearing uncomfortable green rain boots. When I finally returned to the apartment at midnight, my feet were nearly destroyed, and I needed days to recover.

We took all our stuff a few days after Christmas and left Avignon and France for good. Marc and I traveled south by train through Italy to Sicily and Palermo. We spent New Year's Eve 1977 on an overnight ship from Palermo to Tunis, the capital of Tunisia. We arrived in Africa on the morning of New Year's Day. For one week, we toured the country by train and on mopeds. We came to the edge of the Sahara. We ate unusual food like goat's head and a sandwich of sharp red spice. We met and had conversations with many young Arab men. An uneasy tension between Marc and me in our relationship was building. By then, Marc could understand French well – but could not yet speak it. I had a six-month head start over him. I conversed in French with the Arab men we met. Marc could only listen. It was frustrating for him not to be able to intervene in the verbal exchange. He did not like how I spoke with them or what I said. Influenced by feminism, I criticized their macho, sexist remarks about women. Marc thought I was out of my mind to do that, and he was right. I should have been more respectful of the "otherness" of their culture. The young Arab men also told us how much they loved America, and I also made critical comments about that.

When we returned to Sicily, the tension between us was boiling over. We spent a few days in Trapani and took walks on a rocky beach. We separated at the Palermo train station. We agreed that it was not the end of our friendship. We were going our separate ways for now.

I became scared and nervous about being alone again. Yet I met a new travel companion only thirty minutes after separating from Marc. Fate was kind to me. I climbed on a train heading north to Messina and Naples. I entered the six-person compartment. There was a French woman in her early twenties. She was sitting there reading from the "complete works" of the French avant-garde playwright, actor, and theorist of the "theatre of cruelty" Antonin Artaud. I knew a lot about Artaud, so I had a good conversation opening with her. My French was decent enough to converse, even – or especially – about philosophical and intellectual subjects. She was a runaway rebel from her stifling life in Paris. Neither of us had any money, so we joined forces. After we hit the streets in Naples, she immediately found us a place to stay. She was bold. She talked with a group of Iranian medical students. They invited us to crash with our sleeping bags on the floor of their large apartment.

I spent time at the university in Naples. It was February 1977. Something exciting and major was brewing among the youth and students of Italy. The rebellion was crystallizing into solidarity and imminent collective action. The group-in-fusion was forming. You could feel it in the air. It would be a "situationist" protest about the miserable conditions of student life. There were political banners everywhere. They were hanging from the rafters. The halls of the university faculty halls were occupied. There were assemblies, meetings, and passionate debates 24 hours a day. I didn't understand Italian, so I didn't know what the participants were saying.

I traveled further north to Rome and Bologna. I saw the same atmosphere in those cities. The occupation and self-management movement started in the south, in Palermo and Naples, then spread to the universities in the middle of the country, then the north. In Rome, I visited the *Città Universitaria* [university city] campus of *La Sapienza* near the Tiburtina train station. In Bologna, I hung out with the freaks, queers, artists, feminists, anarchists, Metropolitan Indians, autonomists, homeless people, and drug addicts of *Piazza Verdi*, at the heart of the oldest university in the world.

One afternoon, I was walking at the periphery of a student demonstration in Bologna. The military police were stopping all passersby and checking identity cards. I showed one of them my American passport. He asked me some questions, and I replied that I did not understand Italian. Then the *carabiniere* shouted *"Non prendermi in giro!"* ["Don't make fun of me!"] and slapped me across the face. My glasses went flying. I retrieved them from the ground. Fortunately, they were not damaged.

I had little money during my seven months in France and did not eat well. I rarely had the opportunity to cook something to eat in a kitchen. Besides bread, cheese, ham slices, and yogurt, I sometimes bought tomatoes, oranges, and cans of tuna. As a twenty-year-old male, I was stupid about such practical matters and failed to care for myself properly. In Italy, I ate much better. You could get a cheap meal in the student *mensa* [cafeteria] or a family-owned *trattoria* in the university zone. According to the Italian concept of *primo* [first course] and *secondo* [second course], I had a plate of pasta, roast turkey, or a thin steak with a side dish of potatoes, vegetables, a salad, and fruit for dessert. Then a strong *caffè* to not fall asleep.

CHAPTER 8

# Art Students Make Politics: the "Metropolitan Indians" in Italy

In March 1977, art students from the DAMS Arts, Music, and Theatre faculty of the University of Bologna participated in an art, culture, and ideas-inspired political movement called the *Indiani Metropolitani* [Metropolitan Indians] or the fight for *gli emarginati* [the marginalized of society]. The art historian Maurizio Calvesi called the Metropolitan Indians the "mass avant-garde."[108] The ideas and actions of the creative wing of the 1977 uprising were a potential coming to fruition of the utopia envisioned by leading radical artistic movements of the twentieth century, such as Dada and surrealism. It was the possible realization of the dream of overcoming the wall separating art and everyday life, and in the context of new forms of expression enabled by new media and new communications technologies.

What does it mean when a political movement is inspired by art, culture, and ideas? What was compelling for the movement about the symbolism of ethnographic native North American cultures? How does a movement of those excluded by society's economic institutions differ from the Marxist emphasis on the conditions and struggles of the working class?

Another center of activity of this art-political movement of 1977 was Rome. The term *cacciata di Lama* [llama hunt] originated on February 17, 1977. Luciano Lama, the leader of the largest trade union in Italy – the CGIL linked with the Italian Communist Party – gave a speech at La Sapienza University. Lama was greeted by Metropolitan Indians and their creative life-size "Lama dummy," which they constructed as a theatre arts project. "Llama hunt" refers playfully to the domesticated South American animal. Lama told the students to give up their occupation of the faculty halls and go back to their classes. The art student Claudio Tosi made the Lama doll from polystyrene, twine, and wood scraps. It was assembled near Lama's speech on a small stage platform. It looked like a garden gnome or a stuffed animal. A ladder held the strawman together. The head had bristled hair and a smoking pipe.

On March 11, a twenty-five-year-old student named Pier Francesco Lorusso was killed in Bologna by *carabinieri* [military police], who intervened during a physical confrontation between far-left and far-right groups. The incident sparked weeks of protest demonstrations by the student movement and a festive utopian atmosphere reminiscent of the 1969 Woodstock music festival in upstate New York or the Paris Communes of 1871 and 1968.

The participants in the movement were artists, designers, and creative people. They dressed up like Native American Indians and painted splendid, colorful murals all over the city. They wrote commentaries and poetry as graffiti. They built papier-mâché theatrical props for performances at meetings and rallies. The movement protested the living conditions of students, the unemployed, the homeless, and workers in the late capitalist society. It protested the collaboration or alliance of the Italian Communist Party with the Christian Democrat government, which began in 1976 (*il compromesso storico [the historical compromise]*). The movement was inspired by feminist, gay and sexual liberation, ecological, and American 1960s counter-cultural ideas.

Part of the largest mural in the Bologna university zone from the student movement of 1977 (Street art mural by the artist Luis Gutiérrez, Photo by Barbara Zoli, Reproduced with permission from the photographer)

What are the lessons to be drawn from the experience of the interface between art and politics that existed in Italy in 1977?

More generally, based on which principles should the artist *per se* engage in politics?

And are the very concepts of both art and politics now obsolete? Do art and politics align with the logic and organization of the capitalist society, which they sometimes claim to contest?

A great Situationist coup of the 1970s was when Gianfranco Sanguinetti, a colleague of Guy Debord, writing under the pseudonym Censor, sent out his book-length text *Truthful Report on the Last Chances to Save Capitalism in Italy* (1975) to many political and financial power elite members of Italian society.[109] Pretending to be a Machiavellian sage of the Italian ruling class, Censor argued that bringing the pseudo-radical Communist Party into the government would be a brilliant strategic move to consolidate the system of the "integrated spectacle." Debord developed the concept of the "integrated spectacle" in his 1988 book *Comments on the Society of the Spectacle*.[110] This later phase of the worldwide spectacle, with branches in every country, combines features of Western free-market big corporation capitalism and Eastern state capitalism-slash-communism. It is the cumulative effect of five principal developments: incessant technological renewal, integration of state and economy, generalized secrecy, unanswerable lies, and ubiquitous new media to enact an eternal present.

The so-called "extra-parliamentary" far left in Italy of the late 1970s failed. Why? First, because the movement did not sufficiently separate or distance itself from violence. The violent acts of the *Brigate Rosse* [Red Brigades] and parts of the Autonomist movement were morally wrong. Their actions alienated major sectors of Italian society. Second, the liberation-emancipation movements of the late 1970s were unsuccessful because the ideas of their leading intellectuals were too Marxist. They believed too much in "the revolution."

The challenge to the hegemony of the Marxist-Leninist organizations in the late 1970s came from the feminists. These women criticized not only the monopolization of power by males in the Leninist groups – but also the notion of believing in a set of political ideas unless they relate explicitly to one's personal, existential situation. There was a search for a balance of reason and emotion – or between the cerebral rationality of politics and poetic expression.

Feminist collectives and small "consciousness-raising" groups emerged in many Italian cities. Their emphases were interpersonal relations, subjectivity, sexuality, and personal autonomy. This style quickly spread to an entire generation of activists. Sit-ins, street festivals, round-the-clock occupations and assemblies, and alternative media experiments (like Radio Alice and A-Radio in Bologna) spread throughout Italy.

Lea Melandri, a feminist writer of the late 1970s, developed a thinking of the "imaginary institution of society" (influenced by Cornelius Castoriadis), analyzing alienation in everyday life, the mythology of romantic love, and the codes of imaginary sexuality.[111] As a culturally reproducing institution, the family is the "original infamy," the model of dual-triangular relations whose imprint in the psyche undercuts the potential realization of passions and lived experience. Specific instances of power are difficult to face because we are conditioned by categorical modes of thinking and "ideal meanings." "Women," writes Melandri, "are coming to realize the profound estrangement and inhumanity of the language, conceptual system, and gestures of politics, including those which promise freedom and humanity."[112]

Bologna was the showcase city of Eurocommunism, the much-touted alleged overcoming of the contradiction between East and West, of communism and capitalism. The Communist Party had been in power in the capital of Emilia-Romagna since the end of World War II. Its economic model demonstrated how prosperous capitalist development can be with a "socialist" apparatus in control. Built around a system of mechanized farming cooperatives and small enterprises, the region around Bologna has become one of the wealthiest in Italy. Production in the machine and tractor industries proceeds on a commission or piece-work basis.

The "historical compromise," granting the Communist Party a share in state power, attempted to integrate workers' demands nationally into processes of capitalist transformation that had already been successful in Bologna for many years. Beneath the veneer of the official propaganda lay the invisible city of close to one hundred thousand students and unemployed who lived ghettoized in the "university zone." They bore the brunt of the government's austerity policies: no housing, no money, and overcrowded dining and medical facilities.

After the death of Francesco Lorusso, a few store windows were smashed during the subsequent spontaneous marches. The city administration responded by launching a campaign of repression. Police used guns and tear gas to break up crowds. Leftist "free" radio stations and printing presses were shut down. Tanks and armored cars were sent to "reoccupy" the university zone. Bologna was in a state of siege, defended by all major newspapers and political parties, from Communist and Socialist to Liberal and Christian Democratic.

Violent conflicts were expected. Surprisingly, the movement mostly overcame the impulse to violence and discovered the politics of "the festival." Threatened by a society whose information channels condemned them unanimously, and with no possibility of gaining any support from the organized working class, the students turned to the resource of *creativity*.

Normal rhythms of time were suspended as individuals went days on end without sleeping, lived intensely in personal relationships, and ate and partied at so many different places in a few days that they forgot where they lived (or that they had no place to live).

Improvised journals and mimeographed bulletins with names like "Wow," "We Want Everything," "Without Family," and "Let's Write on Ourselves" appeared. The elaborate murals, Native American-inspired "totem poles," street theatre, and choreographed dances enacted satirical portrayals of the family, school, culture, and politics as institutions of power and control in the late capitalist society. The graffiti on the urban walls was ironic, desperate, vigorous, and macabre. It was the insurrection of pure signs against the ruling semiotic order of messages and meanings. Graffiti and street art are symbolic rebellions against the dominant mass media's one-way communication format. It was also breaking away from the language of the political Left, with its stereotyped phrases and linguistically pre-fixed "code of anger."

The DAMS arts students made a Chinese dragon, which they paraded during quasi-forbidden demonstrations. The dragon became a creative symbol of the movement.

The DAMS dragon (Photo by Enrico Scuro, Reproduced with permission from the photographer)

A poem on the glass door of the entrance to the Faculty of Economics and Commerce (my translation):

> The caravan will be leaving for a thousand roads
> They're waiting for an order that will come
> Maybe the memory of a splash of water that will reappear
> It will pass through our eyes in silence
>
> But the switchboard says no
> Do you see that pair of glasses between the madman and knowledge?
> It's always nice to take in the scenery
> on the infinite roads that lead to Rome
>
> I want to speak, but I can't
> I want to love, but I can't
> I want to communicate, but I can't
> They've taken life from us
> Let's take it back

Without the constraint of adapting itself to the Marxist pre-established principles of class struggle and the "seizure of power," the student movement of 1977 created a new experimental space. The movement of students, women, gays, and "the marginalized" was a practical demonstration that a similar revolt could take place in other sectors of society. The students' strength lay in their notion of the *"exemplary action,"* intervening spontaneously on a few concrete objectives – rather than with a declared ideology as political groups at the university had always done. Squatting in empty houses and the "self-reduction" of the prices of film tickets and restaurant meals were shared practices.

In 1957, Albert Camus was awarded the Nobel Prize in Literature. Camus gave a lecture at Uppsala University, Sweden, called "Create Dangerously." In the transcript of that lecture, Camus writes that genuine creation allows for no separation or duality between the creative act and political engagement. Creation in our era means to create dangerously. "Any publication is an act," writes Camus, "and that act exposes one to the passions of an age that forgives nothing." Creativity is born from the desire to realize or accomplish. The person gifted with creative potential cannot bring her own life to fruition if she does not mobilize those creative capacities. "Like all freedom, it is a perpetual risk, an exhausting adventure, and this is why people avoid the risk today, as they avoid liberty with its exacting demands."[113]

The so-called "extra-parliamentary Left" in Italy included political groups like *Lotta Continua* [*Continuous Struggle*] and a publishing collective like *Il Manifesto*,

# ART STUDENTS MAKE POLITICS

both of which grew out of the radical movements and events of 1968. Those two organizations published interesting daily newspapers I often read after learning Italian. Yet those constituted political bodies were not the spirit of the movement of 1977. On the contrary, the crisis of those groups, the rebellion against them, ignited the movement. The rejection of their internal hierarchies, stagnant terminology, purported sexism, and their fixed doctrines drove the events of the late 1970s. After returning to America, my Italian experience made it difficult for me to fit in among American leftist intellectuals and associations, who often had beliefs and styles resembling what the Italian movement had left behind.

## Disney World in Bologna

My time in Bologna brought me into contact with Umberto Eco, the philosopher, novelist, and literary theorist who taught at the university's DAMS Drama, Art, and Music School.[114] Umberto Eco brought art and politics together by developing cultural theory, pioneering the media theory study of hyperreality and simulacra, and advancing the knowledge field known as semiotics. In 1975, Eco published the essay "Travels in Hyperreality," making him one of the earliest theorists

Umberto Eco (left) at DAMS during the Bologna student uprising of 1977
(Photo by Enrico Scuro, Reproduced with permission from the photographer)

of simulation and simulacra (before Jean Baudrillard) and an astute observer of American consumer culture, providing detailed descriptions of the tourist venues in the USA which he visited.[115] Eco wrote about American hyperreality in ways parallel to Baudrillard. Semiotics studies meaning-making (making "fake" or simulated meanings) and artefactual symbols and sign processes, including cultural processes independent of language and linguistic sign systems. Umberto Eco authored at least seven books about semiotics.

In the book *Travels in Hyperreality*, Eco travels to America, searching for the paradox of the authentic fake. Eco seeks out cities that imitate a city. Disneyland is realistic AND fantastical. America is hyperreal because it believes that making a copy of something is the ultimate certification of originality. Everywhere in America, there are architectural replications and ambient simulacra. The simulation process transforms so-called "reality" into an inferior version of the imitation. Hyperreality henceforth rules. The copy becomes the model to which the original must answer. The latter pales in its "graphic resolution" compared to the former. Copying becomes essential to capture the alleged authenticity of the original.

The alligators on the banks of the "real" Mississippi River must be coaxed to come out and be photographed. They do not always make an appearance. Their Disneyland animatronic alligator counterparts always cooperate in the performance. Disneyland California's "precision and coherence are to some extent disturbed by the ambitions of Disney World in Florida." Walt Disney World Resort is 150 times larger than Disneyland. It is a vacation and leisure center of golf courses, sprawling hotels, interactive designer experiences, exotic simulated multicultural villages, shopping, water parks, and science fiction futurism. It is the copy of what is already a copy of life and the merger of life with that copy." The "Pirates of the Caribbean" attraction opened at Disneyland in 1967, featuring animatronic characters. "Pirates of the Caribbean" at the Magic Kingdom in Florida opened in 1973. As an entrance building, the fort-like "Caribbean Plaza" substitutes for the Disneyland version's New Orleans mansion.

In the old American West around Utah and Arizona, one can drive many miles in the desert to nowhere and hope that one's car does not break down from the extreme heat. One arrives at an apparent nineteenth-century town that was an artificial stage setting for numerous Hollywood Western films. Umberto Eco visits towns built from nothing: the more intense the drive to imitation, the more allegedly "real" the simulated ambiance becomes. Visit "the Old West at any dedicated theme park: horse and carriage, steam locomotive train, sheriffs and jail, telegraph agent, Bar-b-Q cookout, Indian raids, native American handicrafts on sale. Yet the average American wears jeans that are not very different from those of the cowboys."

## The Events of March

After the liberalizing deregulation of broadcasting in the mid-1970s, free radio stations in Italy, such as *Radio Popolare* [Popular Radio] and *Radio Onda Rossa* [Red Wave Radio], proliferated. They offered alternatives to the monopoly of the state-sponsored RAI network. That was the background for the outsized role that Radio Alice played in the 1977 counter-cultural liberation movement in Bologna. The name Alice was, of course, taken from Lewis Carroll's *Alice's Adventures in Wonderland*. Franco "Bifo" Berardi – the neo-Marxist philosopher and theorist of media and technology – was a Radio Alice activist. Radio Alice was a living practice of peer-to-peer media sharing of grassroots dialogue and culture. The Situationists, existentialism, and experimental theatre profoundly influenced its concept. Everything was allowed to be said into the microphone: announcements of meetings and events, poetry readings, yoga and meditation lessons, on-the-air dramaturgy, social dreaming, declarations of love, and cooking recipes.

*Finalmente il cielo è caduto sulla terra* [Finally, the Sky Has Fallen to the Earth] was the name of a weekly magazine edited by Berardi and friends.[116] After four issues, the police outlawed the publication, raided its office, and arrested most of the editorial staff. The forces of order believed it to be the instigator of an insurrection. In fact, it was a venture of collective theoretical reflection linked to praxis.

At 10 AM on March 11 in Via Mascarella, four hundred members of the right-leaning international Catholic movement "Communion and Liberation" assembled. Some of the participants beat up five medical students. Thirty leftist *compagni* [comrades or companions] (as members of the movement called themselves) rushed to the scene and were met by the intimidation of about one hundred aggressive adversaries. The state *carabinieri* and the local police were called in. Police arbitrarily beat students. One *carabiniere* shot and killed Francesco Lorusso in cold blood. Machine guns were fired. Francesco ran and was hit and collapsed. He arrived at the hospital and was declared immediately dead. Radio Alice broadcasted the news. Everyone felt pain and anger. Spontaneous gatherings coalesced in all university classrooms and halls.

By 5:30 PM on March 11, eight thousand members of the student movement came together to march and make their response to the murderous brutality of the state known. They marched through Via Rizzoli and Via Ugo Bassi, two of the broadest shopping avenues of Bologna. Unfortunately, some *Autonomia* participants (the branch of the movement that wanted an armed struggle against capitalism) sneaked to the sides of the procession and smashed shop windows. A contingent of the larger movement occupied several tracks at the main train station. Clashes with the police continued throughout the evening. House searches and arrests by the authorities continued during the night.

## CHAPTER 8

On the morning of March 12, many *compagni* loaded into buses to head to the national demonstration that was called to take place that afternoon in Rome. Four thousand students who remained in Bologna marched from *Piazza Verdi* to *Piazza Maggiore*. In the early afternoon, news spread that the police were about to attack the university. The students mobilized and erected barricades to prevent the police from entering the university zone. The police shot tear gas continuously at small groups of people that evening and over the following days.

On the evening of March 12, municipal police or *carabinieri* (diverging accounts exist) invaded the office of Radio Alice. They destroyed equipment, made many arrests, and shut down transmission. The confrontation was broadcast live. According to one account, Radio Alice started broadcasting again one month later. According to another account, an attempt was made the next day – on Sunday, March 13 – to resume broadcasting, but it was interfered with by someone else sending out whistling sounds over the same frequency. The free radio station of the Bologna counter-cultural movement broadcasted on and off for a while – then stabilized and remained on the air for a solid two-year run.

In the summer of 1978, I returned to Bologna. My second and longest stay there was twenty months in a single stretch. I lived as a squatter in an unfinished apartment in Via Mascarella, within walking distance of the university zone. I spoke almost no English that entire time. I learned Italian to the point where I could read novels in the language and express myself verbally very well on any abstract subject, but not about concrete things like physical descriptions. My best friend and fellow American, Marc, was also in Bologna. We met once every two weeks to recount our respective adventures. At age twenty-two and far from home, twenty months felt like an eternity, although it was not. I volunteered at the anarchist free radio station called A-Radio in Piazza di Porta San Vitale and had a part-time job teaching English.

On October 12, 1978, the Communist-Party-oriented trade unions of Bologna made a display in *Piazza del Nettuno* (the Neptune statue and encompassing fountain adjacent to *Piazza Maggiore*) of the remains of a public city bus which had been burnt to a skeletal structure presumably by members of Autonomia during one of their battles with the police. The interior of the bus was completely wrecked. The exhibition of the burnt bus symbolized that the citizens of Bologna were fed up with the far-left student movement. The movement was accused of a crime and was presumed guilty. The act was called *teppismo fascista* [fascist hooliganism].

On October 21, 1978, I was with a group of art students from DAMS who manually constructed a nearly full-sized replica of a public city bus from cardboard, cartons, and papier-mâché. The festive atmosphere of working together on something creative was politically significant. We carried the completed cardboard bus to the middle of the town. I started a chant in Italian, inspired by an incident from my childhood involving the Sunday morning theft of bagels and lox at Pierce Camp Birchmont in New Hampshire when I was thirteen years old: *"Non possiamo dire una bugia! Non possiamo dire una bugia! Non abbiamo bruciato l'autobus!"* ["We cannot tell a lie! We cannot tell a lie! We did not burn down the bus!"] We laid our artistic bus down in *Piazza del Nettuno*, where the trade union organization had displayed the original burnt-out bus. It was a symbolic act of asking forgiveness from the people of Bologna, yet also pointing out the absurdity of discrediting the entire student movement by associating it with the act of one group of vandals. The police immediately swept in and destroyed the ironic bus-like artifact.

October 1978 DAMS city bus activist art project (Photo by Giuseppe Cannistrà, Reproduced with permission from the photographer)

## Last Stop Venice

What did my participation in the social rebellion movement of the late 1970s in Naples, Rome, and Bologna personally mean to me? How did it affect my life? Of course, it made me want to spend more time in Italy. Besides that, I think

CHAPTER 8

it had an ambivalent effect on my life, bringing about something positive and negative. On one side, it brought me closer and deeper (in my fascination and love) to what happened in France in May-June 1968, since the "spirit of May" now became something I lived directly in my life. Yet, on the other side, as the years went by, it became more apparent that 1968 was something to be gazed at only in the rear-view mirror. My obsession with 1968, in a certain sense, ended up holding me back, believing passionately in something ephemeral and passé. It was a chimera. It kept me from dealing with the state of the world and my life in other ways.

My first trip to Europe lasted a little less than one year. The last stop was Venice. I was now twenty-one years old. While wandering in the narrow walkways and crossing pedestrian bridges over Venetian canals, I knew that I was heading back imminently to America, that a modest measure of time in Venice was all that stood between me and the arduous reckoning waiting for me at home. I had no idea what I would do in New York. I was not ready to apply for, nor enroll in, any graduate school Ph.D. program. I had no money in any bank account in New York. My parents were not going to give me any money. They were dead set against everything that I was doing. I had two hundred dollars left in my pocket. I figured I could afford to spend two weeks in Venice before resignedly getting onto a train to Luxembourg. I would get on an Icelandic Airlines flight to New York with the second part of my one-year open return ticket. I would arrive in New York with perhaps fifty dollars left. To stretch my meager funds while in Venice, I would sleep in a youth hostel and eat one warm meal daily. I found a bed in a humid basement in Castello, the *sestiere* [district] where few tourists go.

There are no cars. I walk in an enchanted trance with no sense of purpose or destination. I lose myself sauntering into a small *ramo* [small branch path] or *campiello* [small square] with no exit. I traverse the Grand Canal in an inexpensive public *gondola* or *vaporetto* [waterbus]. I go over the Rialto Bridge. I see *Piazza San Marco*, the *basilica* in Byzantine style, *Palazzo Ducale*, and the ninth-century Church of *San Zaccaria*. The Campanile of St. Mark's Church (the bell tower) was last restored in 1514 when it attained its present form. It was rebuilt in 1912 after its collapse on July 14th, 1902, at 9:45 AM.

I contemplate the statue of the eighteenth-century playwright Carlo Goldoni in the *Campo San Bartolomeo*. I visit the island of *San Giorgio Maggiore*. I take an express boat to the *Lido di Venezia* [Venice Lido], which I know from Thomas Mann's novel *Death in Venice*.[117] I spend a day walking on the beach, getting sand in my shoes, and dipping my toes in the water. I peruse the Galleria art museum established by Baron Giorgio Franchetti in the *Ca d'Oro* Gothic palace, which has splendidly crafted features on the Grand Canal. I look at the art of Chagall and Klimt at the Museum of *Ca' Pesaro*. I spend time in the *Gallerie dell'Accademia* museum, seeing famous paintings of the Renaissance masters Bellini, Titian, Tintoretto, Carpaccio, and Giorgione. It is a few years before the future opening of the Peggy Guggenheim Collection. I hang out at the university in and around the *Ca' Foscari* palace, eating a small breakfast of cappuccino and brioche outdoors, reading a book, and conversing in French with Italian law and economics students.

One afternoon, I ate lunch in a *trattoria* and drank a carafe of red house wine. I became somewhat drunk, and my rationality waffled. I saw and heard a street artist playing music elegantly with the movement of his hands on crystal wine glasses set up on a table. The glasses were vibrating at different frequencies.

I walked to the edge of a wide canal and stared deeply into the water. I went down a couple of steps and nearly fell. Several meters down, I saw scattered blue light with an intense illumination at its core. In the center of the light show appeared an open cylindrical passage through which one could swim. I was mesmerized by the inebriated hallucination. It presented the idea that one could pass through this clandestine wormhole, possibly enabling time travel to the past and/or a jump to a distant location in space. There followed a rapid succession of image flashes in my inner eyes. It was as if a contact lens-like device had been implanted that caused hyperreal apparitions in my field of vision:

- Neon signs
- Cowboy Vic (the neon, arm-waving graphic image outside *The Pioneer Club* in Las Vegas since 1951)
- Fremont Street (downtown Las Vegas, the city's heart before the advent of the corporate casino hotels)
- Binion's Horseshoe (old-style riverboat casino on Fremont Street)
- Golden Nugget (built in 1946, one of the oldest "gambling halls" in Las Vegas)
- Spinning roulette wheel
- The white ivory ball lands on number 36

What was on the other side of the tunnel? And what year is it there? Does linear time exist inside the wormhole? The structure of the wormhole was not stable. It was perpetually appearing and disappearing.

In one of my undergraduate physics courses at MIT, I learned about the speculative possible existence of a hybrid natural-artificial wormhole. The black hole hub could be engineered into a machine for time travel to the past. Two different times in history could be connected through an outward flaring "neck" or "throat."

CHAPTER 8

Fremont Street, Las Vegas (Photo by Larry D. Moore, Wikipedia BY-SA 4.0 Creative Commons License)

The Einsteinian general relativistic property of spacetime curvature, providing the basis for the opposition between the wormhole's two "mouths" (contiguous in space yet deferred in time), would enable this technology.

An effective procedure for generating enough useful "negative average energy density" material to time travel between 1977 and the early 1960s would entail bringing the quantum theory "Casimir effect" to bear on the stabilized traversable wormhole. Engineers would set up two large conductive parallel metal plates, divided by an intermediary vacuum, in separate chambers at each mouth of the wormhole. Thanks to the controlled reduction of quantum mechanical effects, the sets of twin plates would serve as a technology for recycling "virtual" subatomic particles popping into and out of existence less randomly in the energy-fluctuating quantum vacuum. The hyper-concentrated electrical field would create a permanently recurring series of matter-antimatter-spacetime ruptures.

Strangely, the past beckoned deep in that water.

A man selling ice cream from a wheeled hand cart roused me from my strange reverie. "You scream! I scream! We all scream – for ice cream!" he shouted in English to his potential customers. There was delicious Italian *gelato* on sale everywhere in Venice.

It was time for my pilgrimage to the old Jewish ghetto, a visit I promised my father I would make. Jews were forced to live in the segregated quarter for three centuries, from the early 1500s to the end of the eighteenth century. At the end of the fifteenth century, many Jews arrived in Venice from Spain. In contemporary times, there are a small number of Jews living in Italy – less than fifty thousand.

As I walked on the main square of the Venetian ghetto in the *Cannaregio sestiere*, I felt at home. I sat on a bench under a tree and meditated deeply. I agreed with Dad that I would communicate with him telepathically from that spot and at that moment. Suddenly, a black cat came close, stared at me for a while, then ran off. The cat knew many things that I did not know.

I wonder if this neighborhood has an informal *schmatta* (Yiddish word for textiles for garments) industry. I need to buy some inexpensive clothes. Where can I get a nice pastrami sandwich around here? I will have to settle for hummus and a falafel sandwich.

Rainer Maria Rilke was a great turn-of-the-century Prague-born Austrian poet who wrote in the German and French languages. Rilke was passionate about Venice. He visited the city ten times during his adult life. He loved to walk endlessly around the city, becoming familiar with every centimeter. Rilke paid particular attention to the Jewish ghetto. "One travels the *Ponte di Rialto*" [Rialto Bridge], he writes, "on past the *Fondaco de' Turchi* [Warehouse of the Turks] and the fish market… disembarks on the narrow, dirty canals… walks through the cramped alleys and black, smoke-filled gateways out onto an empty, spacious square." Walk past *Palazzo Labia*, traverse a small bridge, then go along *Canale di Cannaregio*, and you arrive.[118]

Rilke writes (thanks to Birgit Haustedt): "The Venetians repeatedly reduced the area of the ghetto, so that the families, who multiplied fruitfully in the middle of all their afflictions, were compelled to build their houses upwards, one on the roof of another. And their city, which did not lie on the sea, overflowed slowly into the sky, into another sea, and around the square with the fountain."[119]

**Departure (Fortgehn), by Rainer Maria Rilke** (my translation)

Plötzliches Fortgehn: Draußensein im Grauen
Mit Augen, eingeschmolzen, heiß und weich,
Und nun in das was ist hinauszuschauen

Sudden departure, standing outside in the grey
With eyes melted, hot and soft,
And now, looking out at what is

O nein, das alles ist ja ein Vergleich
Der Strom ist so, damit er dich bedeute,
Und diese Stadt stand auf weil du erscheinst

Oh no, that is all a mere comparison
The current turns to you and has meaning for you,
And this city rose up because you appeared

# CHAPTER 8

> Die Brücken gehn mit Anstand der dich freute
> Gelassen her und hin in deinem Dienst
> Und weil das alles ausgedacht ist nur: dich zu bedeuten
>
> The bridges go nobly and please you
> At peace back and forth in your service
> And all this has been thought up to have meaning for you
>
> Als wären sie – ja was?
> Der Canal Grande
> In seiner großen Zeit und vor dem Brande
>
> As if they were – yes, what?
> The Grand Canal
> In its most glorious time and before the fire
>
> Und plötzlich hört Venedig auf zu sein
>
> And suddenly, Venice ceases to be[120]

It was time for my sad departure, the melancholy end of this first year in Europe. What Venice meant to me as an individual, perhaps to all individuals, was close to believing and acting *as if* something was there only for me – because my presence willed it into existence. This was very comforting. Yet there was a price to be paid for this narcissistic greed: as soon as you turn your eyes, the object of love, desire, and fascination ceases to be.

I was reading more of Albert Camus. In his lyrical essay "Nuptials at Tipasa," written in 1936 at age twenty-three, he writes of a place where he goes recurringly, which is personally "sacred" to him, where he can replenish his energies, a sublime spiritual home to which he can and will always return.[121] I feel this way about the Roslyn Duck Pond and Venice.

For Camus, that spiritual home is near Tipasa, a town on the Mediterranean coast of Algeria, about 70 kilometers west of Algiers. It is the site of remarkable ancient Roman ruins. "To the left of the port, a dry stone stairway leads to the ruins."[122] One encounters the remains of structures belonging to an ancient metropolitan colony of twenty thousand inhabitants: the Great Basilica, the Basilica Alexander, and the Basilica of St. Salsa. One also sees the traces of theatres, baths, and the ancient harbor. A visit to Tipasa renews Camus' sense of the meaning of life. It is a place where "the gods speak." He communes there with the sun, the sea, the blue sky, the flowers, and the heaps of dry stones. Camus feels the balm of reprieve in this kingdom of ruins from his lifelong existential and political "exile." There is a "marriage" of ruins and springtime, of self and world, of the human body and primal elements like heat, salt, and water.

At Tipasa, Camus' heart grows calm. He learns to breathe. His severe task of "becoming what he is," of discovering his "deepest measure," is eased by the eternally recurrent journey. Tipasa is an act of living. For the artist, the time of living is separate from the time of creation. The latter is the act of "giving expression to life." Living to its fullest without the distraction of *mimesis* will make the work of art possible later. "The mind would grow calm, and the body relaxed, savoring the inner silence[…] I sat on a bench, watching the countryside expand with light […] I felt a strange joy in my heart, the special joy that stems from a clear conscience."[123]

I am nothing without the world, and the world is nothing without me. It is the balance between the humanism of creativity and the posthumanism of "taking the side of objects." The silent dialogue between the world and me incubates our love, which I embrace despite the temporariness of my stay on this planet and my awareness of death.

Sixteen years later, in 1952, Camus writes the meditative essay "Return to Tipasa."[124] He visits the ruins next to the sea to "warm himself against the stones." The Roman relics and vestiges are now surrounded by barbed wire. On his initial return to Tipasa after a long passage of time, Camus is alone. He has the sensation of seeing his younger self, or his double, walking ahead of him. The youthful Camus is the original, and the Camus who is now on his walk is the copy or shadow-self. "On the promontory that I used to love, among the wet columns of the ruined temple, I seemed to be walking behind someone whose steps I could still hear on the stone slabs and mosaics but whom I should never again overtake."[125]

On his second return, several years later, Camus discovers the right contact with the strangeness of his pilgrimage site. He learns to listen to the sounds of silence and stillness:

"No sound came from the village. The sea likewise was silent […] Years of wrath and night melted slowly away in this light and this silence. I listened to an almost forgotten sound within myself as if my heart, long stopped, were calmly beginning to beat again."[126]

Within the silence, one could discern the nearly imperceptible sounds that composed it: the resonance of the birds, the lizards, the sea, the rocks, the trees, the rustling bushes, the ancient columns. Each musician "of the world" made his proud contribution to this majestic and shared output of the orchestra of otherness.

I learned that I, too, Alan, had my instrument to play in the collective expression. This role would be enough. I was strong. Camus: "In the middle of winter, I at last discovered that there was in me an invincible summer."[127]

## CHAPTER 8

The day of departure from Venice arrived. I rose at 5 AM. I walked to the Santa Lucia main train station. I got into a train heading westward towards France. I wanted to visit Marseille briefly. Since one of the main accomplishments of my journey was that I had learned French, I wanted to spend a few final days hearing and speaking the language. To reach France's second-largest city by population, I changed trains in Milan, Ventimiglia, and Nice. I slept a few nights in a youth hostel near the old port of Marseille.

I met an amiable guy from California about ten years older than me who prided himself on being a world traveler. He advised me to go and do seasonal work in Alaska. In a short time, one could make a lot of money by "packing tuna in cans." It was tedious but lucrative work. You had no expenses while there and came away with a large wad of cash. He went to do his tuna work in Alaska once a year for three months. It financed the rest of his year of world travel.

I took a train from Marseille to Luxembourg, then flew across the ocean to JFK Airport. My Dad arrived in his Cadillac and drove me back to the big house in Roslyn. I did my laundry in my mother's washer and dryer, then watched a Mets game on TV. Finally, I went to my room for a long sleep.

CHAPTER 9

# Back in America Only to Leave Again

I was twenty-one years old with a bachelor's degree from Cornell, back from my one-year adventure in Europe. I rested at my parents' upper-middle-class suburban house on Long Island. I had a few home-cooked meals provided by my mother, meandered through the wild woods down the old wooden steps behind the backyard, and communed with the friendly and empathetic *canards* at the nearby Rosyln Duck Pond. I had one hundred and fifty dollars left from the original twelve hundred with which I had started my first European wanderings. I had no other money in the world. I emptied my bank account before the trip I began on the day before the bicentennial Fourth of July in 1976.

I had some valuable Irish stamps in a fancy booklet my father's boss, James Ruderman, gave me as a gift. I could sell those. I had a few (not mint condition) Mickey Mantle and Willie Mays baseball cards from my childhood collection that I could sell for an infusion of cash. I still did not know if I wanted to settle and live in Europe or the United States. But now I was here in America and would have to go to some city somewhere. I talked with my father, who grudgingly gave me three hundred dollars. He said it was a loan and stressed that I would have to repay it soon. I would need to get a job and repay him from my first paycheck.

My parents expected me to be a big money-maker. They had me classified and pegged for that. They believed I should have studied civil engineering at MIT. I should have gone to work with my father in his lucrative business firm designing skyscrapers in Manhattan. In lieu of that, I should now go to graduate school in business management, medicine, or law. My self-understanding and identity as a humanities intellectual was a choice or destiny actively opposed by them. Maybe I would go for an official university Ph.D. in literature or history. Maybe I would live as a poverty-stricken independent scholar writing articles and books in some poorly ventilated attic. Whether I would do one or the other mattered not to them. If I chose to live outside the rules of American money-making and "contributing to society" according to my parents' concept, they were not going to support or help me, either financially or emotionally. I would have to make it alone for money and love.

## Yes, I Lived in Boston

I repacked my small backpack, grabbed my sleeping bag, and headed for Boston. I knew that Beantown was one of the most left-liberal cities in the country – I might feel at ease there. I had some familiarity with Boston from living across the Charles River in Cambridge during my two teenage years at MIT. I went to my old dormitory, MacGregor House, and they kindly let me have a single room there for two weeks. From a bulletin board on campus, I found a cheap room in someone's apartment, which I rented for one month.

I walked up the hallowed steps to the main entrance of MIT at 77 Massachusetts Avenue. I went past the magisterial white pillars and into the open space of cavernous acoustics of the vast, stunning Lobby 7 of the elite science and technology university. Suddenly, I ran into a girl I knew from the Roslyn High School Class of 1972 (my graduating class five years earlier). I had never spoken with her before, but we immediately recognized each other. We had been in one or two English and social studies classes together. Since I was two years younger than my graduating high school classmates, I was an outsider with no circle of friends to stay in contact with over the years. My relationship with the many cohorts who knew each other well and would joyfully attend class reunions for many years into the future was only marginal. But now, this young woman from Roslyn stood in the MIT entrance hall I had walked through a thousand times a few years earlier as an academically precocious young man. Her great significance in my psychobiography was that she was Janet's best friend.

Janet – the daughter of the French teacher who died from a heart attack, the girl I was madly in love with from afar, the girl who was my girlfriend for two weeks during the Christmas break of my freshman year at MIT when I was sixteen, the girl in whom I would remain obsessively and hopelessly in love until I finally forgot about her at age thirty-five. There was Janet again face-to-face with me in the surrogate incarnation of her longtime closest friend. But I was very quickly going to mess the whole thing up. Like the disastrous phone call with Janet from Steve's house when he blurted out his nastiness that I supposedly "hated" Janet, it was simultaneously an opportunity to change the course of my life positively yet a giant chance to blow it all up self-destructively.

For twenty minutes, Janet's best friend and I told each other condensed stories about what we had been up to and experienced the past five years. I told her about Cornell, France, and Italy. I talked about the books and philosophers I had read and how they had formed my first conscious thinking. She told me about her college, her studies in linguistics, and her current freelance artistic work. She was smart and nice. I should have forgotten about Janet right then and there and asked her best friend to have a coffee with me.

She was unexpectedly very forthcoming about what Janet felt and thought about me. She brought good tidings that I could never have imagined. It was astoundingly fine news.

> "Janet thinks about you a lot and would love to see you again," she said. "The two weeks you were together were an important and wonderful time for her. She is single now and has no boyfriend, so there's a good chance you could be happy together. She would be happy to hear that you spent a year in France and learned French."

After everything that I lived and suffered, my mind was too messed up, and I could not handle this. My feelings towards other potentially close people were hurt deep inside. I had developed thick, cynical armor. I did not have the refinement to accept the compliments gracefully. A split-off part of my psyche had gone ugly and rogue. A strange sub-text within my interior monologue and outward speech had turned crude and defensive. I thought of myself as a committed feminist and rarely said anything sexist. But, at that moment, something hidden was unmasked in what was one of the most chauvinist and self-damaging moments of my life.

> One thing has always puzzled me," I replied. "Janet is so pretty and beautiful, and you are not." This girl was homely in my despicable still-adolescent estimation. "I don't understand why Janet is friends with you – since you two are absolute opposites.

An expression of total shock and disgust appeared on the woman's face. Without saying one more word, she turned and abruptly walked away. It was the last – direct or indirect – contact I would ever have with the life of the beloved and coveted Janet.

I settled on doing temporary office work in the business world to make some money. Temp employment agencies were plentiful in both Manhattan and Boston. You had to pass a five-minute typing test to get a paid gig for a few days or weeks. Your responsibilities while you were sitting at the workstation at the company where they might send you might include typing, answering phones, and filing papers. You would type from a supervisor's handwritten scrawled notes onto fresh sheets of paper inserted into an IBM Selectric machine. Rotating the black or grey platen knob on the right side of the semi-automatic writing apparatus, you carefully positioned two blank white pages with a messy blue carbon paper sheet between them to get an original and a copy as output. Or, sitting at a computer terminal, you would keypunch onto cards and make tiny chads or coded notations that had no meaning to you. A few years later, the standard temp job would involve operating a dedicated word-processing device.

Most of the applicants in the employment agency's waiting room were female. A woman emerged from the inner office with a clipboard and my filled-out cardboard application to fetch me. She had a smirk on her face after realizing that I was male and an Ivy League graduate. She was skeptical if I could type 60 or 70 words per minute with one word deducted for each error.

I knew how to type well from my practical business course in the eighth grade. My results on the endlessly submitted-to-five-minute test were erratic. Sometimes, I tried to type too fast and made ten or more errors during the allotted time, thus blowing up the test. The well-dressed employee would then escort me, humiliated, out the exit door of their office to the waiting row of elevators. Sometimes, I typed too slowly and did not reach the required words per minute. I failed at the universal capitalist-technological categorical imperative of speed. Then I faced the walk of shame, the quick vertical descent to the lobby, and my return to "pounding the pavement" moneyless on the outside. Sometimes, I was in the typing whiz zone, acing the performance and scaling the words per minute mountain while making minimal errors. I would be heartily congratulated, get a pat on the back, and be accepted onto the team.

They did not even seem to mind that I had an unkempt hippie beard and long hair. I took so many typing tests that I started to see and feel the typewriter keys in my nocturnal dreams. I felt like the proverbial monkey who would eventually reproduce the complete works of William Shakespeare if given enough time arbitrarily "banging away." Having proved myself to be a proficient key-banger and getting told that I was being sent on an assignment, it was crucial to remember to hand in the IRS tax form declaring that your annual income was so minimal that you wanted to have only social security tax deducted from your paycheck and not the federal, state, and local income taxes which would eat away a quarter of what you earned.

The work at the temp jobs was dull. It was everyday life in the modern world. You looked forward to the coffee break, the lunch hour, and the end of the day. You kept your mind on the money while keeping your eyes on the wall. You focused on the thought that, by 5 PM that afternoon, you would be thirty or forty dollars less poor. You could make more money if assigned to the night shift at one of the major law firms. To get hired for that, you had to claim you had "legal experience." You could get "time-and-a-half pay" or even "double pay" if you had to stay and bang away after midnight. Reading the illegible scribbles and cryptic markings of the stressed-out lawyers working long hours was challenging.

I was in my little bedroom near Central Square in Cambridge and woke up one morning feeling sick. I felt extremely tired. Exhausted. The bad feeling did not go away. After a couple of days, it got worse. I had a high fever. I called my mother on the phone. She told me to come home to the big house in Roslyn. After those few summer months, I would never live in the Boston area again. I packed my stuff, went to the South Street Station, took a train to New York City, then took the Long Island Railroad, changing again in Jamaica, Queens. I walked past good old Roslyn High School and up the East Hills of "Country Estates." The next day, I saw a doctor. He took blood tests and said that I had mononucleosis with a touch of hepatitis. He said I would be OK, but it would take a month to recover. I should rest in bed. I was back in my single bed, where I slept during high school. I watched TV and read books. My mother provided me with food. She was good at caring for me when I was sick.

## Yes, I Lived in Manhattan

After I was healthy again, I decided not to return to Boston. I had even less money than before. I still owed my father three hundred dollars. I put on a sports jacket and tie that I found in my closet in my room in the house in Roslyn. I rode with my father in his car to Manhattan at 7:30 in the morning. I was going to look for a full-time job. Five days a week. Forty hours. Nine to five. I had a copy of the Help Wanted Classified Section from the Sunday New York Times. It was many pages. I had no idea what kind of job I would be qualified to do or would have a chance of getting. I did not know where I should look in the alphabetically ordered list of categories of job offers.

I saw that some company was looking for college graduates who knew some French. Several employment agencies had the same job announcement and description. I called. The man at the other end of the line said to come right over. The client was the French American Bank in the Wall Street area. They were looking to hire about fifteen new employees *en masse* for their back office.

When the male employment agent saw me and my hippie look, he laughed heartily but stayed friendly. Since I was a Cornell graduate and knew French, I would, for sure, get the job. But I needed to cut my hair and shave off my beard. He explained that was an absolute requirement. I hesitated for a moment. My hippie look had become integral to my identity during the last few years. It felt like a loss of self to give that up. My hesitation lasted one minute. I had to move forward with my life. I had to get that job and do it for a while. I did not want to spend more time running around looking. OK, I said. He instructed me to come back after I cleaned up my act.

I returned to Roslyn and stood in front of a mirror. I did a first round on my beard with scissors and a second round with an electric razor. I went to a barbershop. They cut off my long locks. I returned to the Frenchie employment agency the next day. My man laughed even more heartily when he saw me clean-cut. "You look like your long-lost half-brother!" he exclaimed. Indeed, I did. He sent me downtown to my hiring interview at the French American Bank. The salary was $145 a week before taxes. Not very much.

An elderly French woman in the personnel department formally interviewed me. She mildly scolded me for being "lost at sea" and not knowing what I would do with my life. For her, that question translated instantly into "How was I going to make money?"

I was going to work in their "Bill of Lading" department. She took me across the street to meet a younger woman who would be my supervisor and direct boss. The second interview went OK, too. They hired me instantly, and I would start the job the following Monday.

One big problem was that the job was boring and meaningless. A second problem was that the salary was a mere pittance. A third problem was that I had no idea where to spend the nights between my Monday to Friday day shifts within a decent travel time to and from the Bank. There was no affordable place to live in Manhattan. Even if there was, I did not know how to find it. Perhaps across the river in Brooklyn, but I knew nothing about that. The rent would, in any case, consume most of my meager salary. There would be nothing left and there would be no chance to save any funds. Sleeping at my parents' house in Roslyn, which had the one advantage of being rent-free, meant commuting three hours a day. That added boredom and tedium to the boredom and tedium.

I kept up the horrible routine for three weeks or fifteen days. I woke at 7:15, got dressed super-fast, shoved some breakfast down my throat, and got into the front passenger seat of my father's car. It was an extension or re-living of the summers when I worked in his office as a teenager. We arrived in midtown Manhattan at 8:30. I went down the subway, rode from midtown to downtown, and was at my Bill of Lading paper-shuffling desk at 9 am.

The job ended at 5 pm. I took the subway back to Fifty-Ninth Street and Lexington Avenue in Midtown and met my father at the garage where his car was parked. We drove back to Long Island through the heavy rush hour traffic and arrived home at 7 pm. My mother had dinner on the table waiting for us.

I was an office clerk and had to read and approve Bills of Lading in French and English. I had to check the correctness of all the details on the form. I had to send the Bills of Lading to the next step in their processing. But first, I made copies of them at the copy machine and filed the copies. I had to do some typing. There was a large and steady volume of transactions, and I had to learn to be fast. For each commercial contract, there was also the insurance certificate and the invoice to be saved in a three-ring binder. If I learned well and became proficient, I might get promoted to Assistant Manager of the Department of Bills of Lading for International Trade. The Bill of Lading acknowledges the receipt of the goods. The paperwork guarantees trust between the exporting seller and the importing buyer. The Bill of Lading is transferable by authenticated endorsement or any signing over or reassignment of the stipulated goods. The Bill of Lading specifies the quantity and identifies the merchandise's particulars.

All the workers in the back office were in one huge room half the size of a football field. About ten meters behind my workstation were several female French typist-translators who had been working next to each other for many years and hated each other. By the end of each day, tempers ran high, insulting each other vulgarly in French. *"Ferme ta gueule!"* ["Shut your trap!"] *"Je m'en moque!"* ["I don't freaking care!"] (*"Je ne te dois rien, connasse!"* ["I don't owe you anything, dumbass!"] *"Espèce de vilaine vache!"* ["You ugly cow!"].

The Bank had hired a dozen recent college graduates to work in the back office. We all knew each other. We were all unhappy at the workplace. During lunch, we talked and complained about how miserable it was there. The lunch hour was my only free time during the entire day. To save money, I brought tuna fish sandwiches from home. The atmosphere in Operations was disciplinary. The "old timers" who had worked there for years were grumpy. On the Thursday at the end of the three weeks, three of my cohorts told me secretly that they would resign from the job the next day. On Friday, we would be given our paychecks in envelopes for the time worked. For three weeks, my check, after taxes, would be about $350. It was enough to pay back my father's loan. I thought things over that night and resolved to quit the job the next day. Although the total situation was what I could not stand, the real deal-breaker was the long commute adding to the forty-hour week. I had zero time to do anything fun or stimulating from Monday morning until Friday evening. As a new way forward, I came up with the idea that I would look for a part-time job.

I was back in the Human Resources department, doing my exit interview with the same French woman who had conducted my entry interview four weeks earlier. I was composed enough to perceive that, in her mentality, she had altogether assimilated American values. Her tone towards me was one of reproach. I was honest with her in explaining why I did not want to continue as an employee of their Bank. Surprisingly, she accepted my reasons as valid. Her disapproving comments were her assessment of the general state of my life and a dark warning prophecy about my future:

> "Alan, you will have to make money! You will need to do something! You may not like THIS job, but you cannot keep saying no to every situation!"
> "Yes, Ma'am. Yes, Madame Director." *Oui, Madame la Directrice.*
> "You have done OK in life so far," she continued. "I admit that. I consider you to be a friend. But heed my warning! Listen to your friend! If you continue your present course, you are heading for catastrophe!"

In a way, I had achieved my goal. I had succeeded. I was so naïve and inexperienced in the adult capitalist American and New York City worlds that I did not know any other way to obtain $350. I took the money and ran. I cashed the check, paid back my father, and had fifty dollars left in my wallet. To boot, I had the valuable lesson of experiencing what a job like that was like: hell.

## The Best University Ever: The "Free Association"

I had no desire to pursue any relationship with any conventional institutional university, especially not in critical social and political theory, which was my main interest. Perhaps I would go for a Ph.D. in a few years, but I did not want to do that now. I was fascinated by the idea of an alternative or "free" university. In the 1970s, there were alternative free universities in New York City. The most notable one was "the Marxist School." Yet I never set foot in that place, not even once. Their version of doctrinaire orthodox Marxism-Leninism was anathema to me. I was much more attracted to the "Free Association," a cultural and educational Center on 20th Street between Fifth and Sixth Avenues in Manhattan. The school's name was taken from a quotation from Karl Marx and Friedrich Engels: "We shall have an association in which the free development of each is the condition for the free development of all."[127] *Free Association, 5 West 20th St., NYC, NY 10011.*

The Center was open seven days a week. It hosted courses, seminars, lectures, discussion groups, research projects, art exhibitions, and literary readings. There were art classes for kids and childcare. There were film showings, parties, forums, panels with audiences, concerts, and theater events. There were comfortable spaces with sofas where people could "hang out" at all daytime hours and into the late evening. There were kitchen facilities.

There was a schedule of twenty-four courses, meeting weekly for two hours each, Monday through Thursday, in two time slots beginning at 6 pm and 8 pm.

There were thirty-two faculty members. The class offerings included a writers' workshop and sessions in film theory, filmmaking, graphic design, handicrafts, mastering videotape, music, and dance.

Stanley Aronowitz, a heavyweight in the left socialist movement who later would become a celebrity sociology professor at the CUNY Graduate Center, was a driving force at the school and taught American working-class history.[128] Stuart Ewen, later a well-known cultural historian, taught about the mass media.[129] Barbara Ehrenreich, later the author of the bestselling book *Nickel and Dimed: On (Not) Getting by in America*, taught socialism and feminism.[130] Judith Young led a consciousness-raising group exploring masculine and feminine gender identities via body movement, improvisation, theater, games, play, and storytelling. There were seminars on philosophy, psychology, racism, sexism, and the disciplinary and socialization functions of the mainstream school and university system. There was an internship-oriented TV research project. There was a course on the news media and propaganda. There were classes in Marxist and anarchist theory and workplace organizing from a libertarian socialist perspective.

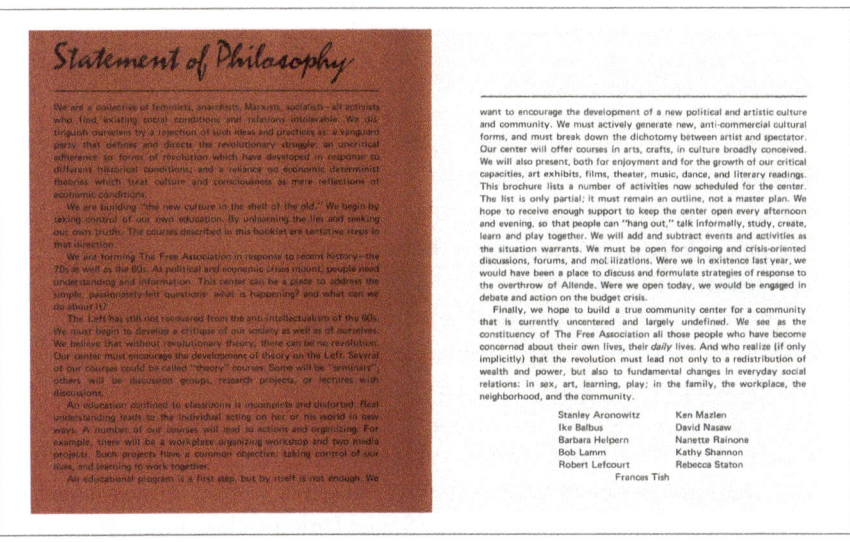

The Free Association Statement of Philosophy (courtesy of Bernie Tuchman, Public Domain)

The Free Association "alternative university" was steered by a collective of eleven people with no hierarchy of rank or power. The collective met once a week to define the direction of the cultural, artistic, political, and educational Center. The project's main goal was to create a community in the sense of an "Antonio

Gramscian" understanding of building a revolutionary popular culture. There was the rejection of the classical Marxist idea of a vanguard party leading the working class and a critique of the assumption that culture and consciousness are reflections of economic relations. Beyond the desired redistribution of wealth and power, daily life experiences must change. This includes how we live in the workplace, the family, and the neighborhood.

In *The Prison Notebooks*, Antonio Gramsci writes:

> Each man, finally, outside his professional activity, carries on some form of intellectual activity, that is, he is a 'philosopher,' an artist, a man of taste, he participates in a particular conception of the world, has a conscious line of moral conduct, and therefore contributes to sustain a conception of the world or to modify it, that is, to bring into being new modes of thought.
>
> The problem of creating a new stratum of intellectuals consists, therefore, in the critical elaboration of the intellectual activity that exists in everyone at a certain degree of development [...] and ensuring that the muscular-nervous effort itself, in so far as it is an element of a general practical activity, which is perpetually innovating the physical and social world, becomes the foundation of a new and integral conception of the world [...] Technical education, closely bound to industrial labor even at the most primitive and unqualified level, must form the basis of the new type of intellectual [...]
>
> The mode of being of the new intellectual can no longer consist in eloquence, which is an exterior and momentary mover of feelings and passions, but in active participation in practical life, as constructor, organizer, 'permanent persuader' and not just a simple orator.[131]

I attended lectures on the anti-authoritarian strands of the revolutionary socialist movement in different periods of twentieth-century European working-class history. Guest speakers who discussed the legacies of anarchism and unorthodox Marxism included the early ecological thinker Murray Bookchin and the Queens College historian Paul Avrich.[132]

My growing interest in anarchism and unorthodox Marxism – when I was in my early twenties – can be explained entirely as having to do with my psychological relation to my parents. I argued with my parents about politics and fundamental questions of economic systems. Their ideas were from the mainstream worldview of the American Cold War era. They dismissed all critical points that I would make about American society or capitalism with the counter-assertion that the economic-social organization of Soviet-Union-style Communism was a terrible totalitarian system. To counterargument their counterargument, I had to find a way to acknowledge the evils of "State Socialism from Above" while asserting that that did not invalidate the idea of socialism since there could alternatively be "socialism from below" with direct democracy. My political philosophy emphasized decentralization rather than autocratic rule from above, whether capitalist or state socialist (the bureaucratic collectivist state). I had to stake out a position from which my parents were seen as right on an essential point but wrong in the big picture. It was the "middle way" of letting them be right without being too right and me being a rebel without going completely off the deep end. I had to achieve individuation.

I met Bernie at the Free Association. He became a lifelong friend. I met Peter, and we became friends.[133] Peter was about my age. He was a philosophy major and very well-read. Peter was a very ethical person who devoted his life to Marxist philosophy and leftist political activism. He was a scholar of the German language. He later became the editor and translator of Rosa Luxemburg's complete works in English. Peter was a principal figure in a Marxist-Humanist organization. After we met, we had intellectual discussions until late in the night.

Peter generously helped me out in my predicament of having no money and no place to live. He invited me to live with him. He had little money and was happy to have someone pay half the rent. Peter lived in a one-room studio apartment on the ground floor of a run-down building on West Twentieth Street, further towards the Hudson River. We had two single beds, one on either side of the only room. I lived there with Peter for nearly one year. The apartment was infested permanently with hundreds of cockroaches. From time to time, the "exterminator" sent by the landlord would come and spray around a bit, and the roaches would be temporarily suppressed. Within twenty-four hours, they would be back in even greater numbers. They were everywhere in the bathroom and the small kitchen corner. It was not advisable to go to the bathroom during the night. When it was dark, the roaches were happily in their element. It was a terrifying sight to see when you turned the light on.

In his early auto-socio-biographical memoir *Down and Out in Paris and London* (1933), George Orwell (pen name of Eric Blair) writes:

> My hotel was called the Hôtel des Trois Moineaux [...] The rooms were small and inveterately dirty, for there was no maid, and Madame F., the *patronne* [female boss], had no time to do any sweeping. The walls were as thin as matchwood, and to hide the cracks, they had been covered with layer after layer of pink paper, which had come loose and housed innumerable bugs. Near the ceiling, long lines of bugs marched all day like columns of soldiers, and at night, they came down ravenously hungry so that one had to get up every few hours and kill them in hecatombs. Sometimes, when the bugs got too bad, one used to burn sulfur and drive them into the next room, whereupon the lodger next door would retort by having *his* room sulfured and drive the bugs back. It was a dirty place but homelike, for Madame F. and her husband were good sorts.[134]

## Stockboy at Macy's

I needed to make some money to survive, to pay my share of the rent, which was $110 a month, and to buy food. I mulled over my options. I was not ready to apply to social science or humanities graduate school. I dreaded the prospect of any full-time job, having to sit in some office somewhere from Monday morning to Friday evening for months. The full-time job scenario was eliminated from consideration. The idea of nine-to-five office temporary work was slightly less oppressive. I experienced that industry in Boston and would indeed do so again later. The big negative of that lifestyle was that you had to constantly hustle – go to interviews, make phone calls to the agencies, experience assignments unexpectedly aborted, face uncertainty – to get the next reliably paying gig. The temp scenario was out. What I needed was a "permanent" part-time job. It would provide steadiness and some income while leaving me a "free man" to do the "independent scholar" reading of books and writing, which is what I wanted to do with my time. The problem, however, was that – in the American and New York City economies as set up – there were very few part-time jobs, and they paid miserably little.

My roommate Peter, whom I had first met at the Free Association alternative university, worked part-time as a stockboy at Macy's. Macy's is the largest department store company and retail chain in America. Its flagship store is at Herald's Square in Manhattan, just north of the Madison Square Garden sports arena. The massive temple of consumerism takes up almost the entire block between Thirty-Fourth and Thirty-Fifth Streets, looking south to north, and between Seventh Avenue and Broadway, viewed west to east. As a non-Leninist Marxist-Humanist and non-violent revolutionary, Peter identified his mission as being part of the proletariat. Despite having his university degree, embedding himself organically in a situation with "the workers" felt right to him. It was very admirable. Peter could get me a part-time stockboy position at Macy's. I went to an interview in a tiny cell among a beehive of cubicles and was hired. The interviewer was astonished that a Cornell graduate wanted this job.

For eight months, I worked at Macy's on a five-day-a-week schedule, four hours a day in the late afternoon. I walked to work from our studio apartment in the Chelsea neighborhood. I had a locker in a changing area where I kept my stockboy uniform jacket. I punched in and out of the mechanical time clock system at the beginning and end of my shift. I inserted my piece of cardboard into a slot and heard the timestamp get loudly imprinted onto the paper-media recording of my name and day and hours worked as input to data processing. After taxes, my paycheck was fifty dollars a week or two hundred a month. I gave Peter my share of the rent and had almost one hundred dollars left for food. There was no money for anything else.

One time, I saw two tardy employees break the glass over the timer in the clock-in machine and move the clock backward with their fingers so they would not be recorded as late.

I was assigned to work in the fabrics department for a few months. It was a widely unpopular post among the large pool of stockboys. No one envied me. My boss was the punctilious and somewhat nasty Miss Lido (her real last name). The department was a large open area of many retail apparel display tables on castor wheels. They were arranged so that each table had four adjacent tables, one on each of its four sides, with spaces for customers to walk around and shop. There were dozens of bundles of fabric set out comfortably apart from each other on each table. There were more organized racks of fabric bordering the walls.

The primarily female customers browsed and rummaged through the fabrics. When they found a fabric they wanted, they would take it to a cutter who would cut one or several yards of material for them. When finding or not finding something they wanted, the customers left the fabrics in total disarray. The fabrics were in all possible disorderly conditions on the table, draped over other fabrics, dangling at the table's edge and on the floor. My job was to go around all the tables during my four-hour shift and restore fabric order. I rolled up each fabric neatly around its fifty-four-inch stand. I stood the textile and board upright again on their proper table in their correct position. As the finishing touch, I made a special kind of triangular fold at the most visible end of the cloth where the first impression on the customer is made. It was very repetitious work. The authoritarian Miss Lido watched me closely.

After my time in the fabrics department, I had a spell as a member of the privileged freelance pool of stockboys. I got to sit around for hours in a back room waiting to be called – to fill in somewhere when some department needed extra help. We were not allowed to read books or listen to the radio but were allowed to look at magazines and comic books. We snuck out to an Irish bar across the street and made sure not to be noticed when we returned.

For a few weeks, I sorted clothes hangers in Macy's sub-sub-sub-basement. It was a place one could hardly imagine existed – a vast unfinished cellar with poor lighting. I was afraid to wander any distance from my work site for fear of encountering rats or even worse. Large white canvas bins of hangers of several sizes and styles had to be separated into output bins where all those of the same type would reunite.

Before I quit Macy's, my final role was in the television department. This was a coveted assignment because you got tips in cash. You would cart or carry the TV just purchased by the customer to his car or a taxi. He would then hand you a dollar or two. You even got to watch TV – America's favorite activity – while waiting for the subsequent request for transport.

Peter told me that he was present when several of the supervisors "informed" a gathered group of stockboys where they could "discretely" pay for prostitutes at an "apartment" a block away, which, they said, "all the supervisors frequent." Peter never took up the offer.

In *Down and Out in Paris and London*, George Orwell writes:

> He led me down a winding staircase into a narrow passage, deep underground and so low that I had to stoop in places. It was stiflingly hot and very dark, with only dim, yellow bulbs several yards apart. There seemed to be miles of dark labyrinthine passages – actually, I suppose, a few hundred yards in all – that reminded one queerly of the lower decks of a liner; there were the same heat and cramped space and warm reek of food, and a humming, whirring noise (it came from the kitchen furnaces) just like the whir of engines [...]
>
> Then the *chef du personnel* [head of personnel] took me to a tiny underground den – a cellar below a cellar, as it were – where there was a sink and some gas ovens. It was too low for me to stand upright, and the temperature was perhaps 110 degrees Fahrenheit. [He] explained that my job was to fetch meals for the higher hotel employees, who ate in a small dining room above, clean their rooms, and wash their crockery [...]
>
> Our cafeteria was a murky cellar measuring twenty feet by seven by eight high and so crowded with coffee urns, bread cutters, and the like that one could hardly move without banging against something. It was lighted by one dim electric bulb and four or five gas-fires that sent out a fierce red breath.[135]

## Trip to Atlantic City

In 1976, voters in New Jersey passed a referendum approving the legalization of casino gambling in Atlantic City. Resorts International Casino – the first legal casino in the eastern half of the United States – opened in 1978. The legalization of casino gambling in Atlantic City was an example of the explosion of gambling social institutions in America in the 1970s. In addition to thoroughbred and harness racetracks, dog tracks, wagering on sports events, illegal "number rackets," sports pools and cards, bingo, *jai alai*, and off-track betting, America was now inundated with State-run lotteries and number games, private-sector sweepstakes, contests, lucky number drawings, gas station games, raffles, corporate giveaways, free vacation drawings, bottle-cap prizes, and newspaper contests. This was the new "gambling society."

Most of the sociological literature on gambling has approached gambling as pathological or deviant behavior. These studies have posited a dichotomy between so-called compulsive gambling (people who ruin their lives) and "normal" gambling (an allegedly harmless passion, pastime, or leisure activity). Does this dualism make sense?

The academic literature has mainly reflected the debate in American society about gambling's morality and legalization, centered around the question of whether the availability of legal gambling leads to a greater incidence of excessive and self-destructive gambling. "Compulsive gambling" is measured by the amount of money lost over a given period of time or by certain adverse social outcomes associated with heavy gambling, such as destruction of family life, job failure, suicide, bankruptcy, unpayable debts, etc.

In 1979, the Congressional Committee on National Gambling Policy commissioned a group of social researchers to undertake a study that produced a 600-page report on gambling activities and behavior in the United States. Responding to a questionnaire, most people said that their major reasons for gambling in casinos were "to have a good time" or "for excitement." The researchers concluded that "gambling is essentially a consumer commodity which people purchase because they enjoy it, rather than because they expect to make money."[136]

In 1978, I took a bus trip to Atlantic City with my brother, aunt, and uncle from Sheepshead Bay, Brooklyn, and my grandfather Samuel Morrison. Three of my four Jewish grandparents were immigrants born in Austria or Ukraine who arrived in New York around 1920. My mother's father was born in Philadelphia in 1896. He had a small self-employed business in midtown Manhattan, where he wrote management brochures for a small group of business clients and self-help pamphlets for the public. I was with my extended family on my way to my first foray into a gambling casino.

We read a book on basic blackjack strategy out loud during the two-and-a-half-hour bus ride. You get two cards. You see one of the dealer's two cards. Always stick when you have seventeen or more. Always hit when you have eight or fewer (except when you might split your cards). Twelve through sixteen is a lousy hand. It is the grey zone where you either hit or stick depending on what the dealer has. With a twelve, you stick versus a four, five, or six and otherwise hit. With a thirteen through sixteen, you stick versus a two through six and otherwise hit. Always double down on eleven except against an ace. Double down a ten against anything except a ten or an ace. Double down a nine against a three, four, five, or six. Always split aces or eights. Never split fours, fives, or tens. Double down a "soft" thirteen through eighteen (an ace with a two through seven) versus a four, five, or six. Split nines except versus a seven, ten, or ace. Split twos or threes versus a four through seven. Split sixes or sevens versus a two through six. Got all that?

I played blackjack for four hours at Caesar's Park Place Casino on the famed Atlantic City boardwalk and won $172.50 (one hundred and seventy-two dollars and fifty cents). It was more than three weekly Macy's paychecks. I sat with my brother at a five-dollar minimum table (my brother lost for the day). I had brought seventy-five with me to blow. I started off betting five dollars per hand. I hovered around even for a while but then got on a winning streak. I increased my bets. The high point was when I bet $15, got a blackjack, and then bet the $22.50 winnings on the next hand and won that too. I was ahead at least one hundred dollars at that point. I had a big shit-eating grin on my face. "Now I'm really going for it!" I blurted out to the other players at the table. I placed twenty-five dollars within the small betting circle on the green velveteen felt table before me. The dealer, a clean-cut, mustached young man about my age, looked at me scornfully.

> Do you think that I've never seen a bet for twenty-five dollars before? Do you think that is a lot of money? Do you have any money? People in here drop twenty-five dollars like it was five cents! "I hope your big winning streak lasts!" he added sarcastically.

My streak lasted until I was ahead by over two hundred dollars. Then I started losing some of it back. I stopped playing and got out of there. I even gave the dealer a gratuity.

## I Hold Down a Full-Time Job for Twenty Weeks

There was nothing in America that made any sense for me to do. I had no profession. I had no desire to pursue any specific line of practical work. It was all "alienated labor" to me, a phrase I had picked up from reading texts of the early Karl Marx. In capitalism, you sell your time and effort to a system run by bosses in exchange for money and survival. You are estranged from the process and the product of your labor. The chain of overseers in the power hierarchy dictates what you must do in your daily activity and how you must do it.

There was also no acceptable way forward for me in any academic career. I was an intellectual and loved the humanities, but universities were bureaucratic and ideationally restrictive. They turned the philosophical, critical social theory, and literary ideas I loved into commodities and business. Universities had hierarchies of power and money, who got a say in things, and what ideas were allowed to be articulated and pursued. Universities, even and especially in the humanities, were where ideas, creativity, and originality went to die.

I wanted to go back to Europe. This time to Italy, not France. To Bologna. The radical left student movement was happening there. I had already tasted it. I experienced it. I wanted to participate in the cultural scene of *compagni* and *compagne* [male and female comrades or companions] and for that movement to shape my thinking. I would have great times, fun times, and meet like-minded young people. Italy was a beautiful country. Italian was a beautiful language. I wanted to learn Italian. The cities and towns of Italy were full of amazing architecture and art. The food markets were endlessly fascinating, and who did not love Italian food? There were lovely women there. The weather in Italy was warm and sunny. There were breathtaking beaches. There were unparalleled landscapes and coastlines. There was ancient and medieval history everywhere. There were archaeological sites. My goal this time was to last in Europe for two years instead of the one year I had reached on my first wanderings. I was like an astronaut who would increase his number of orbits or his time in outer space on each subsequent flight.

I needed to save up some money to make my second departure. I walked into an employment agency. They sent me to an office clerical job at the Computer Center of the City University of New York. The position paid $225 a week, a significant upgrade from my pay at the French American Bank. If I lived frugally, I could save almost all that income and buy my wad of traveler's checks with it. It was a nine-to-five five-days-a-week hell of a routine. I knew that I would not be able to stand that grind for very long, but I had a strong motivation: my dream to leave the country. The woman manager to whom I was assigned as her administrative assistant was authoritarian and nasty. However, shortly after I started the job, she took an extended leave of absence for health reasons. I felt bad for her, but it was good fortune for me.

For several weeks, I sat alone all day at my desk with no one bothering me. I answered the phones. I did some typing and document filing, which was given to me by another administrative services director. I read some novels and the *New York Post* and *Daily News* sports sections. When 5 pm came around, I celebrated being a few dollars richer. At the end of twenty weeks, I could no longer stand the regimen of going there daily. I quit. I had accumulated nearly four thousand dollars. It was more money than I had ever seen or dreamed of.

## Yes, I Lived in San Francisco

As I was getting ready to leave for Italy, I met a beautiful Italian woman from Milan in a New York City coffee shop. A mutual friend introduced Paola and me. We lived an intense and "utopian" romance for two weeks. We spent the nights and the weekends together. It was passionate. I felt like I was in love.

We were sitting at the kitchen table of the apartment lent to us by an acquaintance of hers, having a conversation. Paola asked me how old I was. I said twenty-two. Then she asked how old I thought she was. She looked just a few years older than me, so I said twenty-five. Then she said that she was thirty-eight. She was sixteen years older than me! She did not look her age. I was startled and disappointed. I intuited that our relationship was not going to work. I considered breaking it off. But my feelings for her were deep. I momentarily convinced myself that "age is just a number." I decided to continue with the adventure despite the anomaly of our age difference.

The situation was complicated even more because she was about to leave for Central America. She was an Italian high school English teacher on a sabbatical and was traveling for a few months together with her female colleague. It was their dream trip to the United States and the Western Hemisphere. They had made a precise plan for every week and every month. They were staunch feminists and were not going to modify the plan to accommodate a boyfriend met along the way. After their side trip to Belize, Yucatan, and Cancun, they intended to spend a few months in San Francisco. They regarded "the City by the Bay" as the epicenter of what was left of 1960s American counterculture: New Age spirituality, alternative sexual orientations, alternative gender identities, leftist and feminist political radicalism, and literary and musical creativity. That is what they wanted to experience.

Paola said that we could continue our love affair on the West Coast. She persuaded me to fly out there. I intended to go to Italy, not California, but I reluctantly agreed. To make things even more troublesome, after she left New York, I developed intense abdominal pains, which were diagnosed as appendicitis. My appendix was removed. I was in the hospital for a few days. Nonetheless, I soon found myself landing at San Francisco International Airport.

Our love relationship transformed into a disaster when I arrived in Northern California. Paola had asked me to bring her suitcase, which she had stored in a closet in the Manhattan apartment where we had spent our two honeymoon-like weeks. I absent-mindedly misunderstood her description of the piece of luggage and brought the wrong suitcase. I did not see that there were two traveling cases in the closet. In a moment of non-awareness, I clumsily grabbed the valise of her fellow high school teacher traveling companion.

I stayed eight weeks in San Francisco with Paola and then left. We tried to recapture the magic we had felt at the beginning of our liaison. There were amorous moments, but they were always defeated by disharmony and clashes. I finally gave up. I flew back to New York, then shortly after to London (from where I would travel to Paris and then Bologna). I blew a thousand bucks on my San Francisco detour and started my second European wanderings with only three thousand clams in traveler's cheques left.

Despite my broken heart and depressed misery at things having gone sour with the dream woman Paola, I managed to do some things in and around San Francisco that I wanted to do. For the second time, I hung out for several days at the University of California at Berkeley campus, symbolizing the New Left student movement of the 1960s. I spent many evenings at jazz clubs between the

Mission District, Nob Hill, and Pacific Heights, listening intently to improvisational music that "touched my soul." I went to baseball games of the San Francisco Giants and Oakland Athletics. I had fantastic dinners in Chinatown, at North Beach (Italian) and Fisherman's Wharf (seafood). I played frisbee and played with wildly energetic dogs in Golden Gate Park. I felt the vibes of the hippies and the reverberations of the 1967 "Summer of Love" in the Haight-Ashbury district. I went to readings at the famed City Lights Bookstore associated with the beat poets Lawrence Ferlinghetti and Allen Ginsberg.[137]

CHAPTER 10

# Two Years in Bologna

## The Chimes of Big Ben

After spending my first night in a youth hostel, I went on a pilgrimage to several of London's anarchist and radical left bookstores, the names and addresses of which I had from a directory I found in a pamphlet. At those bookshops, I saw many English-language Situationist literature that was unavailable in America. I browsed and bought a few items. At Housman's Bookshop at 5 Caledonian Road in Kings Cross, I engaged in an interesting political and intellectual conversation with a young man who said he was part of an anarchist collective. He knew a lot about the Autonomist and Metropolitan Indian movements in Italy. He had read French theorists like Foucault, Baudrillard, and Gilles Deleuze. He invited me to sleep in an empty bedroom in the communal house where he and his comrades lived. All the house residents had jobs and were away all day, even in the evening, so I never got to talk much with them. The rent-free "crash pad" gave me a home base to explore London.

I saw Buckingham Palace, Westminster Abbey, Parliament Square, and St. Paul's Cathedral. I visited the British Museum, the National Gallery, and the Tate Gallery. I strolled through Hyde Park and Kensington Gardens. I saw the Big Ben Clock Tower.

Big Ben reminded me vividly of my second favorite TV show after *Star Trek*. I watched all seventeen episodes of the British ITC show *The Prisoner* when they were broadcast in New York on the Channel 13 Public Broadcasting Service station. The show was enlightening.

In the two-and-a-half minute opening sequence shown at the beginning of almost all the episodes, with inspiring music playing, the character initially without a name who soon will be called Number Six, played by Patrick McGoohan, drives past Big Ben in his S2 Lotus Seven sports car that he built himself, on his way to resign in anger from his job as a James Bond-like secret service agent of

British Intelligence. Every episode of *The Prisoner* except for "Living in Harmony," "The Girl Who Was Death," and "Fall Out" shows this opening credit sequence, which many consider the greatest in television history.

After resigning, Number Six returns home and packs his suitcase with travel brochures (and other items). The powers that rule the world refuse to let him leave as a free man. He falls faint after breathing in a knockout gas released through the front door keyhole into the interior of his abode. The authorities *take him as a hostage*. They abduct him and imprison him in The Village. He wakes up in an apartment, an exact duplicate of a portion of his London dwelling. In each episode, he seeks to find out the true nature of The Village and tries to escape.

The residents of The Village have lost their names and humanity and have become numbers. Any escape attempt is futile. The Village is a joint venture financed and operated by both sides in the Cold War. The state apparatuses hold hostage former secret agents and high-ranking military officers of the Western and Eastern blocs. In the first episode, "Arrival," the female taxi driver who speaks to Number Six briefly in French (because "it is international") tells him it is common in The Village to meet Poles and Czechs. In the episode "The Chimes of Big Ben," Nadia tells Number Six that she is Estonian, yet she has an ethnic Russian name and speaks perfect Russian (there is a Russian minority in Estonia). The Village is a facility shared by the mega-state – the "one unified system" – that runs both sides of the Cold War.

Portmeirion, Wales, UK, the filming location for The Village in the TV show "The Prisoner" (Photo by Mike McBey, Wikipedia Creative Commons BY 2.0 License)

The Village is the ultimate in the indoctrination and subjection of the individual. The Establishment has taken over entirely. Individual freedom is dead. People no longer want to think for themselves, even if they are capable of doing so. The Prisoner of the title is the one man who is resisting. Or is he?

The character of Number Six questions the rules of this unified system – with his desire to "go on vacation." The rulers of The Village tell him, in effect: "No, you cannot do that. You cannot simply drop out. It is not allowed." Not only are both sides part of one single system, but if you choose to leave one side, then it must be the case that you have "gone over" to the other side, to the supposed "enemy," even if you have not. The fiction of the enemy must, at all costs, be maintained. You will be held hostage to preserve the illusion of the antagonism. You must stay in The Village until you make up a convincing story "that satisfies us."

In "Arrival," the elderly ex-Admiral is seated at his usual position in front of a chessboard at an outdoor table in the patio area overlooking the stationary boat that never leaves its docking station and goes nowhere. He says wryly to the female character Number Nine (Number Six turned upside down), who has just betrayed Number Six by promising him a supposed escape from The Village via helicopter that was, in fact, a ploy by the authorities to "break him" psychologically: "We're all pawns, my dear."

The Village and *The Prisoner* are about power and control over the citizenry via the media. McGoohan is connected to McLuhan – Marshall McLuhan the Canadian founder of worldwide media theory. All residents of The Village are under constant video surveillance. There is television, radio, telephone, and the daily newspaper, the *Tally Ho*.

The episode of *The Prisoner* "A. B. and C," first broadcast in the United Kingdom on October 13th, 1967, features a futuristic technology of a connection between the neurological experience of nighttime dreaming and the ubiquitous display device of media culture that we call *the screen*. The science-fictional technology is a totalitarian Nazi-like police apparatus of torture, interrogation, surveillance, and mind control, and, at the same time, a cyber-consumerist technology prophesizing the Brain-Computer Interface. The instrument converts signals and information back and forth between neurological and digital-multimedia data formats.

The Village Authorities are both Number Six's hostage-takers and those to whom the demands ("we want information") – whose fulfillment would allegedly lead to the release of the hostage – are made. But the demands cannot be met, and hostage Number Six can never be freed.

In the iconic dialogue between Number Six and Number Two that occurs as part of the opening sequence of fifteen of the seventeen episodes, Number Two tells Number Six that what the Village Authorities are interested in obtaining is *information*:

> Number Six: Where am I?
> Number Two: In the Village.
> Six: What do you want?
> Two: Information.
> Six: Whose side are you on?
> Two: That would be *telling*. We want information ... *information* ... *information*!!!
> Six: You won't get it!
> Two: By hook or by crook, we will.

The usual interpretation of this dialogue is that the Village Authorities seek information from Number Six in the sense of a *specific content of information*. They want the information about why Number Six resigned from his position as a Secret Service intelligence agent. To which side in the Cold War did he defect? To which party offering a large sum of money or other rewards did he "sell out"? But when Number Six replies, "You won't get it!" he implies, additionally, a deconstruction or critique of the *very concept of information* as the Village Authorities have defined it. The concept is invalid because *no information exists* that would satisfy the Authorities to the extent that they would then release the hostage.

To the show's viewers, it is clear that Number Six resigned to go on vacation. He was fed up with *the whole thing* and decided to drop out. His motivation was stress or burnout. In "A. B. and C," even the extreme mind-control technique that reaches the deepest levels of his subconscious cannot uncover any other motivation. No matter how many times throughout the series it becomes clear to the Authorities that Number Six had no other motivation beyond his wish to go on holiday, they still refuse to accept that fact. Only an answer that aligns with what they want to hear would satisfy them, and such an answer will never come.

This is reminiscent of the contemporary sociological situation. The state and the consumer society make endless demands of citizens that the latter can never satisfy. We are endlessly in debt financially and can never catch up with the credit system. We can never stop working. We are all hostages of the system that insists that we self-regulate, buy insurance, manage our identity, and micro-administer our money, health, and desires.

The Cold War of the 1950s and 1960s was, in many ways, a simulation, a fake, a deterrence of the populations on both sides from achieving awareness about what was happening. The media-consuming public had to be convinced of the alleged war's "reality." Death and destruction were confined to proxy conflicts like the undeclared war in Vietnam that we watched on TV or virtualized in the permanent threat of the potential nuclear holocaust.

The novelistic and cinematic figure of the M16 British spy James Bond was a perfect symbol of this simulationist *mise-en-scène* or enactment. The currency of the Cold War was espionage, *information*, and the secret agents' clever tricks, heroics, and martial arts skills. Bond was a "one-man army." *He was himself the war.*

Patrick McGoohan intended to critique the myth of the James Bond-like character. A significant example of a Bond figure was John Drake, whom McGoohan played in *Danger Man*, the TV series in which he starred in the early 1960s. McGoohan took the character in non-James-Bond-like directions: Drake did not use firearms or kiss women. Number Six *is* Drake.

At the beginning of "A. B. and C," we see that the terrorism of The Village extends to the upper echelons of the power hierarchy. Number Two is terrorized by Number One, from whom he receives threatening phone calls expressing impatience to see Number Six psychologically broken. Number Fourteen, the female scientist and medical doctor who has developed the technology to steer a patient's dreams and project them onto a screen, is terrorized by Number Two. Two tells Fourteen that the machine will be administered to *her* if the information extraction from Number Six does not succeed. Theirs is a classic interaction between the hyper-aggressive Nazi commanding officer and the sadistic Nazi doctor.

The dream scene to be implanted into Number Six's subconscious mind is a nighttime party in Paris at the home of the wealthy female host Madame Engadine. Number Six is to meet a succession of three persons (the foreign agents A. B. and C) whom he knew in real life and who are all candidates for being someone to whom he "sold out" or who might be able to extract the "information" from him. To put him into a deep sleep where he will experience the manipulable dream state, a drug is slipped into his tea just before bed. The sleeping Number Six is brought to the Doctor's laboratory, injected with a special drug, and hooked up to the machine. His dream is displayed on a TV- or computer-monitor-like screen.

The male spy "A" worked for the British secret service and became a defector. The female spy "B" asks Six about his intentions after resigning. He declines to answer. Fourteen then uses the technology to talk directly to Six in the dream "Virtual Reality," speaking through the mouth of the female foreign agent. Some "bad guys" are going to shoot "B" if Six does not share the "information." He remains indifferent to her pleas to save her life.

The next day, Six follows Fourteen into the woods, where he discovers her secret laboratory. He gets inside the building through an air duct. Entering the room where the two-way audio-visual output of and input to his dreams have taken place, Six understands the machine and devises a plan to turn the procedure against his interrogators. He discharges the third and final drug injection from its syringe and hypodermic needle and replaces it with water. He returns to his residence. He fakes drinking the knockout tea and falling into a deep sleep.

Returning to the Parisian party for the third time, but this time in partial control of the dream, Six has a strange, surreal encounter with a mirror, bringing to our attention the complex and confusing relationship between "reality" and dream, or between self-identity and the mirror image. The mirror on the wall is tilted, and Six adjusts it to an aligned vertical position. A mysterious blonde woman hands him a valuable earring to bet on the Number Six at roulette. "Number Six," she says to him enigmatically, "I'm sure it's your lucky number."

The roulette wheel spins. The winning Number Six comes up. Instead of thirty-five earrings or a large pile of chips, Six receives a large metal key to an unknown door as his winnings. It is one of two matching keys. Its companion is in the possession of Madame Engadine, who appears for a short time to be the foreign agent spy to whom Six will betray his country. Engadine drives him in her car to a castle outside Paris. There, Six will meet with the elegant Frenchwoman's boss. Fourteen, observing the events on the screen in her lab, decides to give the name "D" to Engadine's boss. In the dream-slash-VR, Six enters the castle and finds himself standing on a street at night. Six encounters "D," whose face is hidden under a black mask. Six tears off "D"'s mask forcefully to reveal that he is Number Two.

*Roulette scene in the episode "A. B. and C" of The Prisoner (ITC Entertainment, Academic Fair Use)*

The corporeal Number Two, watching on the screen, is driven to despair by this startling revelation. Number Six has gained full control of the dream state and has rewritten the narrative. Still in the dream and carrying an envelope supposedly containing the "valuable information" vital to British and American national security, Number Six returns to the laboratory. The dream-yet-conscious version of Number Six confronts the dream versions of Numbers Two and Fourteen. He hands the envelope to Number Two, which the latter believes holds the crucial information. Instead, the envelope contains vacation travel brochures of Italy and Greece. "He was going on holiday," says Fourteen. "I wasn't selling out," says Six.

Number Two is psychologically broken. The orange phone emits an ominous alarm sound. Number Two knows that Number One is calling.

## The Beaubourg Effect

After London, I spent two weeks in Paris, sleeping mainly in a youth hostel. I spent some time with Anne, an American friend of a friend who had lived in Paris for a long time. Anne was a social anthropologist who had just published a book about the American model of the "advanced psychiatric society," co-authored with the French sociologist Robert Castel.[138]

I spent the fortnight practicing my spoken French with as many interlocutors as were willing to talk with me. I browsed through literature and philosophy books in bookstores. I took walks in the Jardin du Luxembourg and the Bois de Boulogne. I visited the Shakespeare and Company English-language bookshop, which had many books in all categories and literary novels. I saw the vibrant life in the *Quartier Latin* and along the *Rive Gauche* [Left Bank] of the Seine. I inspected the second-hand book stalls. I explored the boutiques and art galleries on the Île St. Louis and the Île de la Cité. I visited the Café de Flore (on the Boulevard St. Germain), which Albert Camus frequented. I traveled to the Nanterre University campus west of Paris, where the historical student uprising began on March 22, 1968.

The milestone architectural project of the Centre Pompidou in the Beaubourg Quartier of Paris had just been completed eighteen months before, in January 1977. The high-tech building complex dedicated to culture in the fourth arrondissement symbolized the spectacular-commodified marriage of art and technology. Le Beaubourg, as the Parisians called it, housed an extensive library and cultural research facilities. The structure was designed principally by the architects Renzo Piano and Richard Rogers. The edifice gives off the effect of being turned inside out: pipes, ducts, circuits, and infrastructural networks are exposed on the outside. Different functional systems of electrical power, plumbing, heating and air conditioning, and emergency exits are identified according to a color-coding scheme.

While in Paris, I read the cultural critique *The Beaubourg Effect*, which Jean Baudrillard wrote and published in 1977.[139] For Baudrillard, the Beaubourg is a "monument to mass simulation effects," a project claiming to support culture that neutralizes all cultural content, accomplished via the massive, imposing apparatus of its Disneyland- or shopping mall-like universalizing form. Baudrillard does not view the architecture of the Pompidou Center negatively. He identifies a contradiction or critical tension between the ambition of the French government's cultural sector and the secret resistance of the masses of visitors to the official, unthought-through implementation of simulation and hyperreality.

The cultural operation of the Pompidou Center is "implosion." It works "like an incinerator, absorbing and devouring all cultural energy" – both the vast contents exhibited within it and the liveliness of the neighborhood environment around it.

The masses resist hyperreality ironically via acts of hyper-conformity. They are the "silent majority," subtly subverting the powerful institution and its trumpeting of the exalted modernist values of art, knowledge, freedom, and humanity. Few visitors wander among the museum's contents. Tens of thousands of curious spectators move through the interstitial spaces of the enveloping structure, traipsing the tunnel corridors, reclining on the oddly shaped furniture, and sitting on the floor.

The masses, generally speaking, are what sociology, in its perennial search for "the social," can never capture. The masses are non-representable. Surveys and election polls are inaccurate because the respondents say what they believe the social scientists and poll takers expect them to say.

## Learning Italian in Milan

I traveled next to Milan. It was 1978, and there was still one month of summer weather. I resolved to spend the month in "the Fashion Capital of the World" and devote myself every day to learning the Italian language. I had familiar methods for drumming a foreign language into my head, which I developed and refined two years earlier when I learned French. I painfully memorized conjugations of regular and irregular verbs and many details of grammar. I perused some easy-to-read novels. I listened to the radio and watched television. I read daily newspapers. I initiated conversations with people who were willing to talk with me. It was the beginning of two years of speaking almost no English. I immersed myself in Italian and French.

Although my love relationship with Paola blew up in my face in San Francisco, we declared that we would remain friends. Weirdly enough, she gave me the keys to her luxury apartment near Porta Venezia in Milan. I did some cooking in her kitchen. Perhaps unconsciously and semi-intentionally, I had a couple of accidents and ruined some of her culinary equipment. Paola's flat was a superb headquarters and location to perform my autodidactic learning of the Italian language and explore the city in the hot summer weather.

I witnessed vast rallies of political parties in Piazza del Duomo with thousands of attendees. I gazed with admiration at the Museo Teatrale alla Scala opera house. I walked inside the Duomo cathedral, awed by its gorgeous stained-glass windows. I made the arduous climb up the stairs to the roof terraces of the Duomo, taking in the Medieval-era aesthetic and neo-Gothic details of the magnificent structure. The arches, tapered spires, supporting buttresses, and pinnacle crowning ornaments seemed infinite in numbers and beauty.

I went to jazz clubs and young people's cultural centers. I experienced the underground music scene. I strolled along the picturesque Corsa di Porta Ticinese, looking at the canals in the Navigli district. I spent endless hours in the greenery of the Giardini Pubblici [Public Gardens]. I followed the winding paths of the sprawling Parco Sempione. Sitting beside tranquil ponds, I started conversations with young and old people to get language practice.

I was in the business of saving money, so in early September, I hitchhiked from Milan to Bologna instead of taking the train. I walked to the edge of the big city and onto the entrance ramp of the highway. I stuck my thumb out. A young

Alan N. Shapiro hippie freak, age 22 (Alan N. Shapiro private collection)

hippie-looking guy in a van picked me up almost immediately. My stumbling, broken Italian was good enough to converse with him. He was driving to the middle of Bologna, so I thought, great, I've done it. I'm home free.

He was enthusiastic about the leftist youth movement in Bologna. He said I was making a good decision to go there and settle for a while. I would experience something historical and fascinating. But there was one small problem: the guy was a heroin addict. And he was proud of it. I had never had a conversation before with someone on heroin. He had just "shot up." Amazingly, he gave a long speech singing the praises and extolling the virtues of heroin. What was terrifying was that he was at the wheel driving on the highway and kept falling asleep. He had a standard cigarette in his mouth. He went through a continuous cycle. He falls asleep. The cigarette drops from his mouth into his lap. The cigarette burns a hole in his pants at his upper leg, which wakes him up. He places the cigarette back in his mouth. Rinse and repeat. Endlessly. He gave new meaning to the expression "asleep at the wheel." I arrived in Bologna and was relieved to leave the van still alive.

## Twenty Months in Bologna

I lived twenty months in an empty apartment in Via Mascarella, between the vintage and flea markets of the Parco della Montagnola and the Botanical Garden and Herbarium at Via Irnerio. There was a housing shortage in Bologna, and I was fortunate to find any sleeping arrangements. A friend of a friend inherited the apartment but was not yet ready to move in. The shower had just been built, and there was no kitchen. I had a bed, a desk, and a chair in my small room. There was nothing else, not even a closet. I lived with my clothes out of a suitcase. My books and notebooks were piled up on the floor. I was essentially a squatter. The elderly conservative woman who owned the building wanted me to leave, and eventually, I had to go.

Every day in Bologna was an adventure. I walked around its historical center, which has narrow passageways and alleys, many without cars from morning until evening, its small and large public squares, and its architectural *portici* [arcades]. I felt happily lost in a stimulating labyrinth. Everywhere, there were historical monuments and churches. The colonnaded porticoes are constructed in front of houses and buildings. They are an omnipresent feature of the streets. You don't need an umbrella in the rain or a baseball cap to protect yourself from the hot sun in summer. Sometimes, you could see an arrow lodged in the brick or stone, shot by an archer, and stuck in place since the Middle Ages. A few porticoes, such as the one of Isolani Palace in Piazza Maggiore, are from an earlier historical period and were built partly with wood.

Bologna is an avatar of perfect urban planning, with its nine *Porte* [city gates or entrances] and its circular layout streamlining the city's compact size. Bologna is circumscribed by the *viali* [ring roads] avenues aligned with the ancient walls.

Standing on the terrace of any apartment on a higher floor in Bologna, you can see the characteristic ubiquitous terracotta rooftops. One of the city's "nicknames" is *La Turrita*, a reference to its many tall towers. At the high point in Medieval times, there were about 120 erected towers. Some were watchtowers for military purposes. Some were status symbols. In Piazza di Porta Ravegnana, across from the impressive Feltrinelli bookshop, where I browsed frequently, stand the imposing *Due Torri* [twin towers]: *Torre degli Asinelli* and *Torre Garisenda*, named after two rival wealthy families of the twelfth century. More than once, I climbed the narrow staircase of 498 steps to the top of *Torre degli Asinelli*.

A second "nickname" of Bologna is *La Dotta* – the city of the learned and the erudite. Bologna is the home of Europe's oldest university, believed to have been founded between 1088 and 1115 AD. The higher educational institution was famous for reviving the study of Roman law. From 1562 to the beginning of the nineteenth century, the university's center was Palazzo dell'Archiginnasio (in Piazza Galvani), still home today to the municipal public library. In 1803, the university moved for good to the zone between Via Zamboni and Via San Vitale. The rector's office, the faculties of many academic disciplines, and the university library are in Palazzo Poggi.

From the apartment where I was "squatting," I had a five-minute walk to Piazza Giuseppe Verdi, the hub of the social life of the students. For one and a half years, I had a fixed daily routine Monday through Friday. I read a book or wrote for three hours in the morning. Then I walked to the area of Piazza Verdi, went to the student *Mensa* cafeteria, or met friends in a trattoria for lunch. I had a part-time job in the late afternoons teaching English, from 3 to 6 PM, five days a week. After lunch, I drank my strong coffee and socialized some more. I hung out in Piazza Verdi every day. There were artsy cafes, bohemian bars, and gaming arcades. I walked with a sense of strange ambivalence past the Johns Hopkins American University Center – the School of Advanced International Studies in Via Beniamino Andreatta – on my way to the stop of the bus that would take me to my part-time teaching job.

One of the centers of the cultural revolution in Paris in 1968 was the Cinema Odéon in the Latin Quarter. There is a Cinema Odeon near the University in Bologna. I posit the existence of a secret wormhole corridor between these two cinemas: we can time travel back and forth between 1968 and 1977, nurturing, cultivating, and developing our dreams to become reality.

Yet a third "nickname" of Bologna is *La Grassa* – the "fatty town." I became enamored of many typical yet simple unique dishes of Bologna: for example, *spaghetti al ragù* [with meat sauce], *tortellini in brodo* [in broth], *tortellini alla panna* [with cream], *tagliatelle verdi* [green], *punta di vitello al forno* [baked veal], and *lasagne al forno* [baked lasagna].

I sometimes attended lectures and seminars on semiotics (linguistics applied to culture or the study of cultural "signs") taught by the DAMS Arts, Music, and Theatre faculty. The colorful, creative murals and graffiti from the 1977 student uprising were still all over the university zone. The book *Indiani in Città* [Indians in the City] collects photographs of most of the murals.[140]

In the early evenings, I went for my *passeggiata* [ritual stroll]. I continued with the book I was reading while sitting on the grass in the peaceful natural setting of the Giardini Margherita. I drank wine in an *osteria* [tavern] while chatting about politics, life, and the world with friends and acquaintances, improving my Italian and learning more about the Italian way of life. Sometimes in the late evenings, I worked volunteer shifts at the anarchist punk rock nightclub.

Piazza Maggiore is surrounded on four sides by the Basilica di San Petronio church, the Palazzo dei Banchi, the Palazzo dei Podesta, and the Palazzo Comunale. Adjacent to the main square is Piazza del Nettuno and its famous Fonta del Nettuno [Neptune Fountain]. The aura of the monumental bronze Neptune statue is enhanced by the adornments of four mythical cherubs and four sirens on dolphins, which are exquisitely represented. The imposing figure was designed in 1563 by the Flemish sculptor Giambologna. It is a fantastic piece of public art.

To resist the bewitching song of the Sirens, which lures sailors to their doom, Odysseus stopped up the ears of his crew with wax and had them bind him to the ship's mast. He wanted to hear the divine sensuous vocals and soak up the knowledge of "whatever happens on this fruitful earth" that the Sirens' song imparts, without being led astray to his death.

In Piazza Maggiore in the evening, the public social life of the city was in full swing. Large groups of men engaged in lively conversations. In the summer, classic and contemporary Hollywood films were shown in the open air on a large screen. The famous Torre dell'Orologio [clock tower] is in one corner of the Palazzo Comunale. The tower was built in 1444, and its clock has operated since 1773.

I met a French woman from Toulouse named Hélène, who was visiting Bologna. She became my girlfriend for six months. She had a small apartment in Paris. We were always together and went back and forth between Bologna and Paris. My advanced learning of the two Latin-based foreign languages continued. A friend of Hélène was studying sociology at the Parisian graduate school *L'École des hautes études en sciences sociales* [School for Advanced Studies in the Social Sciences]. She was a Brazilian woman named Nancy who was writing a doctoral thesis with Claude Lefort, the libertarian socialist political philosopher whose books I read and whose system of thinking I was very interested in.[141] Nancy went on to write important books in Portuguese about North-South global economic issues and ecological thinking.[142]

I met Lefort personally through Nancy and had coffee and lengthy discussions with him in French. He did not speak any English. Lefort and the Greek thinker Cornelius Castoriadis were the leading figures of the ex-Trotskyist *Socialisme ou Barbarie* intellectual group of the 1950s, a major influence on the 1968 student movement at Nanterre and the Sorbonne.[143] I met Castoriadis and had a long conversation with him.

I went with Hélène to Formentera for one month. Formentera is a small, beautiful, sunny Spanish island in the Mediterranean Sea southeast of Valencia. Together with Ibiza, Mallorca, and Menorca, it is one of the Balearic Islands. It was a thirty-minute boat ride from Ibiza to Formentera. Hélène and I had very little money. We slept on the beach in sleeping bags. Every day was hot summer weather, and it never rained. The nights were peaceful. The star-filled night sky was magnificent. A community of thirty or forty hippies of many different nationalities was sleeping on the unspoiled white sandy beach. There were parties at twilight with music, dancing, storytelling, wine drinking, and pot smoking. During the day, Hélène and I read books, swam in the sea, and took long hikes. When we wanted privacy, we took our sleeping bags to the shelter of the adjacent shrubbery and sparse woods. Some of us who had some money rented cottages. There was a spirit of sharing. We took showers at the cottages of friends and used the kitchen facilities. Once every two days, we spent the equivalent in Spanish pesetas of five dollars per person at the local restaurant. For that price, you could eat a large portion from a platter of delicious paella with rice, vegetables, chicken, and seafood.

Hélène had to return to manage some business affairs in Paris. I loved Formentera and wanted to stay as long as possible. I had some money in my squatter apartment in Bologna. I had some money in Hélène's apartment in Paris. But I had only brought enough money to Formentera for a few weeks. I "rationed" my funds by keeping most of them "locked up" in other locations. Early in the morning on the last possible day, with nearly nothing left, I walked to the small tombolo island of Espalmador to get the ferry to Ibiza. I believed I had just enough cash left to buy a boat ticket to Barcelona. From there, I would hitchhike to Paris. When I got to the ticket window in Ibiza, however, it turned out that I only had enough for a one-way ticket to Alicante, which was way south of Barcelona. Unfortunately, this would add another leg to the hitchhiking I needed to do. Alicante was south of the point on the mainland coast due west of Ibiza and more than five hundred kilometers, or five hours south by car, from Barcelona.

I arrived by boat in the late afternoon in Alicante with fifty cents left in my pocket. I walked to the highway entrance ramp and stuck out my thumb. A Spanish businessman stopped. As luck would have it, he was going all the way to Barcelona and took me there. It was dark when I thanked him and got out of his car. I knew that I could not continue hitchhiking until sunrise. I stayed awake all night. I walked around La Rambla and Plaça Reial. Those were central areas of Barcelona where there were many outdoor cafés. I did not even have the coins to buy something to drink. I could have been arrested for vagrancy, but fortunately, that did not happen. I conversed with some of the prostitutes. I sat down for two restful hours in a chair belonging to a street bar that had already closed.

At dawn, I followed the signs to the north highway and walked forty-five minutes to the entrance ramp. Hours went by, and no driver stopped. I felt desperate. It was hot. I had nothing to drink or eat. I contemplated looking for a phone booth in a post office, calling my parents in New York "reverse the charges," and asking them to wire me money. Finally, a male Spanish driver stopped and took me into his car. But he was only going ten kilometers north. It helped very little practically, but at least it got me away from that hopeless spot where I was standing.

I got out of that vehicle, and another car instantly stopped. I climbed inside. It was a young German heterosexual couple. They spoke English. They had been vacationing in Spain and were driving back to Germany. They would take me to the south of France. They gave me bread, cheese, and fruit to eat, and water to drink. However, they each spoke to me separately and did not speak to each other. There was no three-way conversation. I came to understand that their relationship was in a terminal phase. They fought while on holiday. They broke up. They were no longer on speaking terms. They stayed together only to carry out the practical task of returning to Germany. I cultivated two separate friendships.

We crossed over into France. It was nighttime again. We pulled off the highway to a tranquil spot. We slept outdoors in tall grass in our sleeping bags. In the morning, I was wet from the dew. We ate breakfast. They dropped me off near some toll booths where cars slowed down. They gave me ten French francs, and I was grateful for it. An Arab truck driver took me in. I sat in the passenger seat of his cabin. He drove to Paris. We conversed for several hours. He delivered me to the street of Hélène's apartment. My ordeal was over. I found my stash of cash, went to the supermarket, and bought various foods. In the language of the 1970s, I "pigged out" and then "crashed." I lay down and had a long sleep.

Hélène was in love with me. She would have been happy to stay with me long-term. I made a big mistake breaking up with her. It was typical of the mistakes that I made during that period of my life. Hélène was a fine person and a beautiful woman, both in her inner and outer beauty. Yet I thought she was not intellectual enough for me. That was a dumb idea.

I was back in Bologna. One day, I reached the second anniversary of my departure from New York on my second mission to create a life somewhere in Europe. I was happy to reach that milestone. It was an achievement. I also felt a sense of failure because I knew that the trip was ending soon – and without a definitive outcome. I would have to go back home. I was out of money. The income from my part-time job teaching English was not enough to cover my expenses. I had slowly used up all my traveler's checks, exchanging them from dollars to Italian lire and cashing them out. The three thousand I had at the start of the two years was gone.

My best lifelong friend Marc – my roommate and fellow student of European Intellectual History at Cornell – was in Bologna during the same years as me. Marc felt very much at home in Bologna and Italy and decided to stay permanently. As much as I loved many things about Italy, I sensed and decided I could not live there. It isn't easy to pinpoint why, but I can think of two ways of expressing it. First, Italy felt too different from America. Second, Italy felt in conflict with my individualist nature or desire to live anonymously with respect to *all cultures*. Italy seemed to have a "strong" culture. To live there, I would have to, so to speak, "become Italian." Speak with the simultaneous body language of an Italian. Relate to my daily life, immediate physical surroundings, and objects of possession like an Italian. Be very social and outgoing like an Italian. Cook like an Italian. Returning to the first way of describing it, I did not want to live in a country so different from America that staying long-term would make me homesick. I would need to find a country with enough similarities to America not to miss my original homeland. It would have to be a country with a "weak culture."

I closed my affairs in Bologna. The day arrived when I packed my stuff and headed to the train station. Sadly, it was time to go back to New York and America. But first, I would end the two-year trip just as I ended my first one-year European voyage: *Last Stop Venice*.

CHAPTER 11

# Las Vegas in Venice

## Winning at the Venice Lido Casino

I saved the $172.50 I won in the casino in Atlantic City in a special envelope. I was going to use it for gambling the next time that I entered a casino. I took my "big winnings" for my next round of European wanderings. In Venice, the idea of going to a casino came up again. I was staying with a male acquaintance of mine, a native of Venice. Tony was a friend of my close friend Marc S. in Bologna. He had a humid, steamy ground-floor apartment in one of the *sestieri* districts of the city, where tourists do not venture. He and his longstanding girlfriend separated a few months before, and now he had a new girlfriend. I cultivated a Platonic friendship with his old girlfriend Francesca.

One evening, the four of us were sitting around in the dampness. I told them the story of my winning trip to Atlantic City. I embellished the story by proclaiming that I had developed a system for winning at blackjack. I explained that most players concede a substantial statistical advantage to the house by making haphazard decisions in the basic strategy of whether to "hit" or "stick" after receiving their first two cards and seeing the dealer's face card. By making the mathematically correct decision in every possible situation (including knowing when to double, split, take insurance, or surrender), one could reduce the house advantage to a statistically negligible minimum.

There was an elegant casino at the Venice Lido called the Casinò Municipale di Venezia. My friends knew about the casino but had never been inside. Tony was a former hippie and vagabond like me, but he had settled down and ran a moderately successful carpentry business. To my surprise, Tony expressed his belief that I was destined to win a lot of money at blackjack. He announced that he intended to "stake me" money to play. He would give me four hundred thousand lire (about $480), and I would start playing the next night. If I lost, it would be his loss to absorb. If I won, we would split the winnings fifty-fifty.

I immediately assented to Tony's proposition. I thought he was perhaps ingenuous. He was mesmerized by what he perceived as my American ingenuity in coming up with an infallible blackjack system. For me, it was a no-lose scheme. If I lost, the losses would not be mine. It did not seem important to think about what losing might do to our friendship, nor how it might feel to win and then hand over half of my winnings to my "backer."

I had never been inside a European casino. In America, you can enter a casino wearing unlaced sneakers, Bermuda shorts, a Hawaiian shirt, and a beanie cap. I left my beanie cap at home in Roslyn. In Europe, there are attire requirements – men must wear dark shoes, a dinner jacket, and a tie. I had no such accouterments with me. Tony was happy to lend me a complete monkey suit. The shoes were several sizes too big, but the garb fit me OK. I knotted my necktie. Tony's erstwhile girlfriend and I got onto the *vaporetto* that steamed us to the Piazza San Marco. From there, we took the fancy Casino Express Boat to the Lido.

Casino Express Boat to the Lido (Photo by Alan N. Shapiro)

It was ten o'clock in the evening when we disembarked at the landing dock of the genteel ludic setting. The other passengers were casino regulars who knew exactly where they were going. We followed closely behind. My palms were sweaty. My throat was dry. My heart was beating rapidly. My breathing was labored. The pit of my stomach was lumpy. My asshole felt on fire. The lobby had marble statues.

There was a crystal case filled with gilded jewelry. The red embroidered carpeting slipped under our feet and guided us up a plush staircase. We waited in line to show our passports or identity cards to a tuxedo-clad clerk and purchase our admission cards. The casino official noticed that Francesca was a high school teacher, an *impiegato dello stato* [a state employee]. She was, according to Italian law, prohibited from entering a casino.

Francesca was seriously disappointed to discover that she would not be allowed in. It was also a letdown for me – I had counted on having a woman at my side for luck. "Luck be a lady tonight." New arrangements would have to be made. Francesca said she noticed an outdoor bar fifty meters to the left of the casino, facing from the dock – and that she would wait for me there. It was July. The weather was balmy. I thought it would be boring for her to "hang out" at the bar. I offered to gamble for only one hour and to meet her at an appointed time. She said I should not concern myself with her or how she would pass the time. I should direct all my thoughts to the game. The only criterion for how long I play should be what is required to achieve my goal. *L'importante è che vinci* [what matters is that you win], she said. She kissed me on the cheek, squeezed my hand, and walked toward the elevator doors just as they opened. "Good luck!" She mimed it with her lips, smiled, and disappeared. Now I was alone, a cultural exile in a foreign land, wearing my fine zoot suit and cloddish shoes, just me versus the tables. Nothing to lose, everything to gain. I handed my admission card to the next tuxedo man and walked with feigned assurance into the sumptuary interior.

The action was in full swing. The sights and sounds stirred my excitement further. I resolved to observe for a while before starting to play. A half-dozen roulette tables took up the main salon of the establishment. There were other European games that I did not know, like *Chemin de Fer* and *Trente et Quarante*. In the backroom a couple of baccarat games were played. There were two blackjack tables. I asked someone why there were only two blackjack tables. He explained that blackjack was an American game and had only been recently introduced at the *Casinò di Venezia* as an experiment.

I spent quite a while studying events at the roulette tables. I went to a cashier's window and bought four hundred thousand lire in chips.

I had never played roulette before. I placed a twenty-thousand-lire chip (about twenty dollars) on black. Black was my color. The white ivory ball landed on number eleven. Black. I won. My first roulette win – ever. A sacred moment. One of the croupiers placed a chip on top of the chip I wagered. "Let it ride, Alan," I heard a voice from a mysterious "elsewhere" speak into my ears. The ball went around and around the wheel and landed on thirty-five. I won again. I left the now eighty thousand lire in the rectangular space for black. The ball came to

## CHAPTER 11

rest in the number two. Where once were twenty thousand lire, there were now one hundred and sixty thousand lire. It was a handsome stack of chips. I scooped it up. I figured three wins in a row were lucky enough. I switched from betting on black to betting on "odd," designated in European roulette with the French word "impair." Odd was my friend in the arithmetic world of even and odd. I laid a twenty-thousand-chip down in that box. Number eleven came up again. Impair. Odd. I let the initial stake plus the winning chip ride. Number twenty-one was the winner.

I doubled the previous bet. As the ball spun around frenziedly, I sang in a low, melodious voice: "Impair! Impair! Impair!" The ball flirted with the eight and the ten. For a split second, I was sure I lost. Then it bounced back up and came to rest peacefully in the slot of twenty-three. I won six gambits in a row! I was ahead two hundred and eighty thousand lire! I discreetly removed my pile of chips, stuffed them into the right-side pocket of my sports jacket, and headed for the two blackjack tables in the back room.

There were seven seats for players at each table. Both tables were full. Other casino visitors were waiting for seats, standing behind the players in the game, and sometimes placing chips on the "bet behind" option space. I was in an informal queue. Everyone was aware of who would get the next available free place. Finally, a seat to play blackjack opened for me. I sat down and started to play, wagering the "table minimum" of ten thousand lire (about ten dollars) per hand. I did well for a while. I was winning more rounds than I was losing.

One key difference between blackjack in European versus American casinos is how the player signals if he or she wants to hit or stick. In America, this is done (primarily) with hand gestures. You either wave your hand across (horizontally to the table) for "no" or tap the table with an index finger or knuckles or fingertips for "yes." In the Venice Lido casino, it was instead done with speech. You said "carte" if you wanted a card or "reste" if you wanted to stick. I observed something peculiar. I noticed that the European blackjack players habitually and repeatedly made a fundamentally wrong decision. They played way too conservatively. Almost all of them would always stick when they had a 12, 13, 14, 15, or 16, regardless of what card the dealer showed. They were terrified of "busting" or going over (losing instantly). This was an incorrect play. According to the mathematical odds, one should hit despite having a poor initial hand, where one risks going over if the dealer has a card favorable to him, like a 9 or 10.

One time, I had a 16, and the dealer had a 10 showing. I hit, got a picture card, and went over. Two or three of the other players advised me that I must always stick with a 16. After a similar sequence occurred a second time – I had a 16, the dealer had an ace, I asked for another card and again went over – they became genuinely angry and started cursing me in Italian! Not only did they disagree with my decisions, but they believed that my moves were messing up the flow of the cards for them!

In my struggling yet comprehensible Italian, I stated firmly to them that my decisions were entirely my own. I wasn't going to be influenced by them, so they might as well shut up about it. My luck changed, and I won several hands in succession. The next time I had a 16 and asked for a card, I got a 5 for 21. The next three players stayed pat. The dealer pulled a 10 to his sixteen and went bust. *"Bravo! Lei è bravissimo!"* ["Good! You are very good!"] they exclaimed. We all increased our bets, and it happened again. I hit on another 16 and got a 4 for 20. The dealer went bust again. *"Lei è un genio! Lei è furbo, molto in gamba!* ["You're a genius! You're clever, really with it!"] I was either an idiot or a genius, depending on the final result.

After a while, a new dealer sat down. This one was barely familiar with the proceedings of the game of blackjack and was literally learning on the job. I noticed that he was proceeding very fast from left to right (from his vantage point) in noting, then acting upon, the decision made by each player to hit, stick, double, or split, then moving on to handle the decision of the next player. Since all the other players besides myself never took another card when they had a 12 through 16, he assumed that each player with a hand in that count range would stick. The dealer moved on to the next player with his index finger without waiting for the verbal utterance of "carte" or "reste" by the preceding player. He took for granted that players with a certain total could not possibly want another card. According to the rules, this was incorrect behavior on the dealer's part.

I hatched a scheme to take advantage of this unorthodox circumstance. I could afford to increase my bet since I was more than six hundred dollars ahead. I started to play one hundred dollars per hand. I waited until I had a hand in the 12 to 16 range, the dealer had a high card, and the player after me had two cards of 11 or less, so he was going to hit for sure. The moment arrived. The dealer had a 10 showing. I had a 12. The player to my left had a low hand and was bound to hit. As expected, the dealer glossed over my "free agent" right to take a card. In a split second, he automatically gave the next player the next card from the multiple decks in the "shoe" device. I saw that card was a 9. "Wait a minute!" I shouted in Italian. My knowledge of Italian was now going to be worth some money! *Io volevo una carta!* ["I wanted to take a card, and you never gave me the chance!"] The dealer apologized. He admitted his mistake. He took the 9 from the next player and placed it over my 12. I had 21! I won the hand easily. The next player was given the next card. His hand improved as well, and everyone was content.

I repeated the trick two more times. It was not the kind of thing one could do often. But it was worth three winning hands or three hundred thousand lire. I had so many chips stuffing my jacket and pants pockets that I lost count of how much I was ahead. Knowing that my luck could at any time turn south and I would start losing it all back, I got up from my chair. I tossed a gratuity at the croupier and declared my seat free. I slipped quietly away. Within a few seconds, another body replaced mine, and everyone at the table forgot that I existed. I recovered my anonymity and slipped back into the crowds in the main roulette salon.

I went to the cashier window near the front entrance. My sight was now fixed on that entrance, which I focused on as my exit. I started emptying my pockets and unloading my chips onto the counter. The dignified and finely dressed male employee counted them and arranged them in neat little heaps. *Lei è andato molto bene!* he exclaimed. I will never forget those words. [*It went very well for you!*] He counted out a large collection of hundred thousand lire notes and placed them in front of me. I inserted them into my wallet and went to the open air of the

The closed and boarded-up Lido Casino, many years later (Photo by Alan N. Shapiro)

splendid summer. I found Francesca at the outdoor bar where we agreed to meet. With an extravagant gesture, I showed her the money. She was elated but not surprised. We hugged and kissed and drank champagne. Two men with whom she had been chatting were flabbergasted by the bank notes and hearing the story of my big win. We took the Casino Express boat back to Piazza San Marco. We went for a celebratory lobster dinner in a fancy restaurant.

I gave Tony half of the winnings, as we agreed. For the next three afternoons in succession, I repeated the routine. In the late afternoon, I got dressed in my monkey suit. Francesca accompanied me. We hightailed it on the express boat to the Lido. The casino opened at 3 pm. I won about a thousand dollars daily, mixing between roulette and blackjack. The fifth day was the day of reckoning. I lost a thousand and never set foot in that casino again.

If I wager on red or black in European roulette, *pair* or *impair*, *manque* or *passe* *(the numbers one through eighteen, or nineteen through thirty-six)*, I have a nearly even chance of victory or defeat, of gaining an amount equal to my stake, or of sacrificing the money that I have set down, taking into account the house advantage that the thirty-seventh number, the zero, affords to the gaming establishment. In American roulette, there is a zero and a double-zero, thus thirty-eight numbers.

In both European and American roulette, the *En Prison* or *La Partage* rule is sometimes operative, meaning that when you make an even-money bet and zero (or double-zero) comes up, you get back half of your wager. You might get this half back right away, or whether you get the full wager back or not will be decided by the next spin of the wheel. The nonpositive number(s) on the wheel of chance is/are not red and not black. An employee positions my half-lost chips onto a narrow line between further acquisition and forfeiture. The issue is deferred.

The two outcomes, being up or down, getting ahead or falling back, kicking ass or getting kicked, winning a bundle or crapping out, steamrolling or biting the dust, are two separate and distinct stations of existence into one of which I cross over following the croupier's throw and my instantaneous visual recognition of into which compartment the ball has fallen.

The wheel or dish-like device is spinning rapidly. The small metallic orb goes 'round and 'round the track inclined towards the middle of the board. The ball gets deflected by a metal stud. It collides into several ridges. If the white ball tumbles into the slot of a black number, I taste the rush of triumph and easy street. Otherwise, I taste the bitterness of hard knocks. The tiny sphere bounces up from the first pocket with which it flirts and lands disadvantageously.

The two results, winning and losing, and the differing circumstances which they respectively bring about, are seemingly divided one from another. But this is only appearance. There is a system, or a shared basis, to which both winning and losing belong. It precedes them and makes them both possible. It is a system of participation – call it obsessional neurosis or addiction – or call it a seductive play to which I assent. I consent to have my mood, my emotional or psychological state, suddenly affected by a change in fortune or an exterior event.

Parallel to the pairing between gain and loss, there are intimate couplings: pleasure and pain, love and hate, sado and maso, yin and yang. A gambler who comprehends the underlying commonality between winning and losing might achieve sovereign indifference towards the value of money and intuit the secret flow of the game itself.

The four days of winning a thousand dollars per day, minus the thousand dollars lost on the fifth day, divided by half since Tony – my "backer" – got fifty percent of the winnings – my net profit was fifteen hundred dollars. This may not seem like a lot, but it was a massive sum for me. That gambling win in the casino changed my life, at least for the next few months. Or, dare I say, forever. After two years of living in Bologna (and part-time in Paris), my funds were down to nearly zero, and I was slated to get on a flight back to New York. Instead, the fifteen hundred dollars enabled me to postpone my dreaded return to America.

That moolah financed a four-month trip that would take me to Trieste, through ex-Yugoslavia, to several Greek Islands, and then behind the Soviet-ruled totalitarian Iron Curtain to Budapest, Prague, and East Berlin. Then, a week in Frankfurt, Germany, followed by my happily postponed flight to New York.

I decided to experience the warm sun, beaches, clear sea water, delicious exotic food, idyllic beauty, pretty villages, caves and crevices, breathtaking clifftops, cliffsides, and sunsets of some of the Greek islands. I also wanted to see and learn something about Eastern European countries still under Communist rule. It was 1980. I made a plan to travel by train through Yugoslavia, stopping in Zagreb and Belgrade on the way to Athens.

I would start the trip in Trieste, the north-easternmost Italian city that bordered Slovenia. Trieste interested me because the Irish James Joyce, one of the greatest writers of the twentieth century, lived there as a young man for many years. Joyce befriended the literary writer Italo Svevo, whose novels *La Coscienza di Zeno* (*Zeno's Conscience*), *Una Vita* (*A Life*), and *Senilità* (*Emilio's Carnival*) I recently found rather easy to read in Italian.[144] I wanted to see the historical and elegant Triestine stock (commodity) exchange building, which plays a significant role in Svevo's famous novel earlier translated as *Confessions of Zeno*. Svevo claimed that all of James Joyce's major literary works were conceived or came to maturity in Trieste.

For my final two nights in Venice, I rented a room in the Liassidi Palace Hotel. On my last full day, towards evening, I saw the mysterious wormhole again. I sensed its capabilities for physical space-jumping and time-traveling. I drank wine. I was very susceptible to the effects of alcohol. The Italians say (in Latin): *In vino veritas*. Was my sighting of the wormhole "real" or hallucinatory? I had a male Italian friend whose favorite (slang) word was *allucinante*. He said it every two minutes, reacting to everything. I believe it meant *awful, horrible, terrifying*. In my mind, *allucinante* resembled the English *hallucination*. There was something significant about that word or concept concerning my fate.

In that twilight hour, after watching children frolic in a playground on *Rio dei Greci*, near the *Ponte dei Greci* [the bridge of the Greeks] in the Castello district of Venice, near the *Zorzi Liassidi* palace on the *Rio San Lorenzo* canal, not far from the famous *San Giorgio dei Greci* church with its masterpieces of Byzantine art, I discerned the throat and mouth of the wormhole. I saw rapidly circulating undulations several meters down in the water, circumscribing the same wild light show I observed in astonishment the first time I was in Venice three years earlier. There were blue spiraling resonance waves. There was dissipated light at the periphery of the singularity and luminescence at its core. Looking closely, I saw the extreme astrophysical phenomenon of the rip in the fabric of spacetime.

Peering through the open cylinder at the center of the compact region disrupting classical Newtonian physics, I made out distinct images of old-time Las Vegas, circa the early 1960s, the "city without clocks." High-resolution visions of places flowed by in quick succession. Glittering neon signs – those fluorescent tubes or lamps glowing as gas fills them and charges them with high voltage. There it was, visible: I saw the Las Vegas Strip on Las Vegas Boulevard: Castaways, Caesar's Palace, the New Frontier, the Silver Slipper, the Dunes, the Stardust, the Desert Inn, the Tropicana, the Flamingo, the Sahara, the Sands, and the Riviera.

I saw Fremont Street, downtown Las Vegas, in the early 1960s at night: Binion's Horseshoe, the Four Queens, California Club, Pioneer Club, the Coin Castle Casino, the Golden Nugget, and Del Webb's Mint – which Hunter S. Thompson made famous in his 1971 journalistic memoir-novel *Fear and Loathing in Las Vegas*.[145]

I saw the comedians Milton Berle, George Burns, and Red Skelton entertaining on stage. I saw the celebrities Bobby Vinton and Mitzi Gaynor singing and dancing in front of audiences and fans who adored and idolized them.

I saw the cars on the road of that time: Ford Mustang, Buick Riviera, Pontiac GTO, Plymouth Barracuda, and Chevrolet Corvette, Corvair, and Impala.

I saw endless rows of five-cent mechanical slot machines in Las Vegas casinos. These machines had spinning reels and a lever on the right side to be pulled.

I saw a blackjack table where the players were dealt two cards face down and held them in their hands while making subtle, furtive gestures.

CHAPTER 11

I saw a roulette table where the numbers to the left and right of the zero were two and twenty-eight, not twenty-six and thirty-two.

I returned my eyes from what I saw through the wormhole and focused my sight on the immediate physical surroundings. "I am in Venice; I am in Venice." I recited it like a mantra. Across the *Rio San Lorenzo* canal and next to the *Rio San Lorenzo* pedestrian bridge, I saw the large block lettering of the MOKA EFTI CRAZY BAR or *bàcaro* [Venetian tavern], home to fine-quality Colombian coffee and banter about soccer and politics.

In the wink of an eye, I lost my rationality. I jumped into the water of the canal in search of the wormhole. It was madness. I was fully clothed, including my shoes. I held my breath, opened my eyes underwater, and dove forcefully several meters down. I looked for it. There was nothing there. It disappeared. I came back up to the surface. Fortunately, no one saw my bizarre action. I climbed out. I was soaking wet. I regained my senses. I immediately realized the absurd thing I had done in the craziness of the moment. Of course, no wormhole was "really" there. There was no secret wormhole corridor between the two cities of

Near the "Crazy Bar," I hallucinated seeing the wormhole deep in the water of the canal (Photo by Alan N. Shapiro)

Venice and Las Vegas. Yet, for one moment, I mistakenly believed it might be there, making the crucial error of believing too literally in the narrative of my life, journey, and destiny that I was permanently creatively writing. Ashamed of myself, I went to my hotel room to change my clothes. I showered. I dried my shoes on the balcony. I took the clothes I had been wearing to a laundromat. The canal water was certainly not clean.

It was a minor psychotic episode. With everything I had lived since childhood, I realized there was a small streak of madness within me. It was a lack of being on the ground, a sort of "flying." I would need years of psychotherapy to regain the grounding and be certain, in a self-confident and healthy way, that such a psychotic moment would never happen again. I needed both the meaningful freedom of expression of my creativity and a firm earth beneath me.

## Yugoslavia, Greece, Hungary, Czechoslovakia

I wanted to spend a chunk of time on some Greek Islands, then briefly observe life under Communism in Budapest and Prague before flying back to New York just as my money would run out. Perhaps it was typically American to set out on a whirlwind tour where I would visit four European countries in four months. At any rate, one fine morning, I left my Venetian friends and traveled east to bond with my imaginary literary friends James Joyce and Ettore Schmitz (actual name of the pseudonymous Italo Svevo).

After a few days in Trieste, I got on a train to Zagreb, curious to experience something of a society that called itself socialist but with an alternative decentralized model. The "President for Life" of the Socialist Federal Republic of Yugoslavia, Josip Broz Tito, had just passed away on May 4, 1980. The country was mourning, and there was some political turmoil and uncertainty. There was a constitutionally mandated transfer of power away from the federal government and towards the six constituent republics and two autonomous provinces.

I was unsure what to think of the Yugoslav model of socialism. On the one side, after the Tito-Stalin split of 1948, Yugoslavia allowed greater individual freedom in some areas (freedom of artistic expression and freedom of religion). It implemented "socialist self-management" with the workplace democracy of workers' councils in factories and companies. On the other side, there was an intense cult of personality around the autocratic Tito.

I planned initially to stay in Zagreb for only a few nights. I slept in a lower bunk of a double-decker bed in the youth hostel. I noticed a casino for international businessmen and travelers in Zagreb on the top floor of the Hotel Intercontinental. I was still so bitten by the gambling bug that I decided to go there one afternoon. I did not have a jacket and tie, but the casino lent me one. I was the only player in the entire casino. I asked to play blackjack. They knew the game well. I played one-on-one against a dealer, won two hundred dollars (they played in U.S. dollars), and left quickly.

Thanks to a chance encounter, I stayed in Zagreb for one month. While standing on a platform waiting for a tram, I asked a young man for directions, and we got into a conversation (in English). He took me to a café, and we got to know each other. He was a film and sociology student. His name was Adrian. One of his parents was German, and he had lived part of his life in Frankfurt. I then met some of his circle of friends. One of them was a professor of the sociology of culture with whom I engaged in a lengthy and deep intellectual discussion. The professor owned an empty apartment and said I could stay there for a few weeks. Suddenly, I had an attractive place to stay for no money.

Another of Adrian's friends was a young woman who was a philosophy student. She was smart and pretty, and we had a romantic relationship. She was conservative in some of her values, and there was no sex. The focus of her philosophy studies was reading the works of Karl Marx. We had disagreements in our respective Marx interpretations, which dampened our budding love relationship.

During the day, I hung around the university, visiting the film and theatre departments of the Academy of Dramatic Art. I attended seminars and lectures. I met Eric Bentley, who was there giving talks on Bertolt Brecht's theatre arts.[146] Bentley was a prominent British-born American playwright and the primary editor and translator of Brecht's works in English.

Adrian took me to meet his grandfather, whom he said was one of the finest twentieth-century poets in the Croatian language. The grandfather was ill and bedridden. He explained that he had been a friend of the great British and American poet W.H. Auden.[147] Adrian's grandmother cooked a fine dinner.

> As far as I could tell, there were no vegetables in Communist Zagreb. Once a day, I ate a meal in a café, cafeteria, or inexpensive restaurant. The lunch or dinner always consisted of meat (beef, pork, or chicken) and potatoes (usually French fries). Perhaps there were a few leaves of lettuce but no tomatoes. After weeks of no vegetables, I complained to a Croatian acquaintance. "It is indeed unusual to want to eat vegetables," he replied. "That is not part of our culture. But no, it cannot be that there are no vegetables in Zagreb! I will take you to a fine restaurant in a luxury hotel for tourists, and they will surely have vegetables!"
>
> At the luxury hotel, we sat outside. I told the attentive waiter I would like to have a portion of vegetables with my beef goulash. "Of course, we have very fine vegetables!" A few minutes later, he returned with a lovely platter of peas, carrots, and canned mushrooms.

I traveled by train from Zagreb to Athens, making a brief stop in the Yugoslav capital city of Belgrade. In Athens, it was easy to hook up with any one of many international groups of hippies whose destination was one island or another. I blended in with a swarm going to the island of Ios in the Cyclades islands in the Aegean Sea. I stayed on Ios for two weeks. I drank a lot of wine and smoked a lot of pot, which fogged my mind. I have little recollection of details of what happened to me on Ios. The food was excellent. I had a lovely small room. The late summer weather was delightful. The white sand was soft under my feet. The water was crystal clear at Mylopotas Beach. There was rock music and dancing at the nightclubs. For hiking, there was a nature reserve and rocky landscapes. I met a Swedish girl with whom I had a romantic affair. She was a university student, but I cannot remember what she studied.

I befriended a Dutch guy who wanted to travel to the remote island of Amorgos, the easternmost member of the Cyclades group. I accompanied him. There was only a boat once a week debarking and taking passengers at Amorgos, so choosing that landing place was a serious commitment to stay there for a while. The island has a saying: "Welcome to Amorgos, nobody will find you here." I stayed there for two weeks. It was a reinvigorating time of solitude, reading novels, swimming, lying on the beach, and walking in the rugged rocky mountainous terrain on winding hiking trails. I sat for hours drinking Greek coffee and reading a book in a traditional village with no tourists. I talked a lot about life with my Dutch friend, who was trying to get off his drug habit. I naively constructed a romantic

Ios, Greece, 1980, age 24 (Alan N. Shapiro private collection)

narrative in my mind that I was in the middle of nowhere, far away from all civilization. Then, the owners of the second small hotel-restaurant where I stayed (for five dollars a night) told me they also owned a Greek diner in Hoboken, New Jersey. It's a small world, after all.

After returning to the Greek mainland, I took an overnight boat from Thessaloniki to Dubrovnik (the "Pearl of the Adriatic") in southern Croatia (Yugoslavia). I slept in the open air in my sleeping bag on the ship's deck. During the night, we passed close to the Albanian coastline. Dubrovnik's impressive city harbor, medieval architecture, fortified and well-preserved city walls, and picturesque Old Town were sights to behold in the mid-morning light.

To travel from Dubrovnik to my next destination, Budapest, Hungary, I had to undertake three legs of a journey. First, I took a three-and-a-half-hour bus to Split, the largest city by population on the Croatian coast. Second, from Split, it was a six-and-a-half-hours train ride to Zagreb. Finally, I climbed on the overnight train from Zagreb to Budapest.

During the late 1970s, I read the quarterly academic journal of neo-Marxist "critical social theory" called *Telos*.[148] The journal, which appeared in a book-like format, emanated from Washington University in St. Louis under the supervision of editor-in-chief Paul Piccone.[149] *Telos* sought to develop a coherent theoretical framework for the New Left student movement on American college campuses that began in the late 1960s. The journal was paramount in introducing key twentieth-century European luminaries such as the "Frankfurt School" critical theorists, utopian thinkers like Ernst Bloch, and analysts of "everyday life" like Henri Lefebvre to an American audience.[150]

Some of the editorial board members of *Telos* had contacts with dissidents in Eastern Europe. Andrew Arato, a professor of sociology at the New School for Social Research in New York City and himself born in Budapest, gave me the addresses of two Hungarian anti-government intellectuals whom I could contact.[151] The situation in Hungary in 1980 was relatively liberal. The individuals I met were unafraid to have open discussions, even sitting in a café with a Westerner like me. The university lecturer and the journalist with whom Arato put me in contact both loved to talk. They pontificated in long discourses.

Something about Budapest that engraved itself in my mind during the few days that I was there is that several times, I saw bodies of dead persons lying on the sidewalk. The recently deceased seemed to be homeless and alcoholic. They were ignored for a while by passersby.

The sights of the heart of Budapest on the banks of the River Danube, especially in the evening, were unbelievably beautiful. The Buda Castle Quarter, the

Matthias Church, Fisherman's Bastion, the Hungarian Parliament Building, St. Stephen's Basilica, Vigadó Concert Hall, and the Danube Chain Bridge were breathtaking experiences.

The train ride from Budapest to Prague took about seven hours. I traveled overnight and arrived in the early morning. Compared to Budapest's social-political atmosphere, Prague was much more deeply entrenched in the system of Soviet totalitarianism and bore the brunt of its full force. The first thing I observed in the immediate area of the train station was large groups of Russian schoolchildren visiting this crown jewel satellite city of their glorious Empire. They wore school uniforms, waved Russian flags, and sang Soviet Russian patriotic songs.

Every tourist was required to exchange a certain amount of Western currency (U.S. dollars) into Czech *koruna* [crowns]) times the number of days of his or her visit to Czechoslovakia. All the foreigners queued up at the train station's exchange window. The state employee asked me how many days I intended to stay in the country. I replied: "Five days." He then told me how many dollars I should hand over to him. The amount of Czech currency that he gave me was a minuscule sum. I could not afford a hotel room or any decent meals.

A few hours later, I was walking on a deserted street, trying to figure out what to do. I heard an insistent questioning voice behind me, saying in English: "Are you Jewish? Are you Jewish?" I turned around and said yes. It was an elderly man. He held out a large wad of Koruna banknotes. He was offering a so-called "black market" exchange. I gave him two U.S. twenty-dollar bills. The exchange rate he gave me was twenty times that of the official rate. I was rich! With all that Czech currency, I could live like a king for the next five days!

Later, when I met some other Americans, they told me that there were stories in the newspapers about several American and British nationals who were serving one-year prison sentences for exchanging money on the black market. Sometimes, the person offering to exchange money with you is a government agent entrapping you into an illegal transaction.

I got a very comfortable room in a luxury hotel. I ate sumptuous meals in the fancy restaurants of other hotels. I got a big surprise at the end of my five-day stay. At checkout time, a team of hotel clerks handed me my bill. I was relaxed, knowing I had enough cash to cover it. But they wanted to see the state-issued document verifying that I obtained the money at the official exchange bureau. They knew that I did not have the document, and the team had a good, sinister laugh. Oddly enough, they were happy to take some of my American Express and Barclay's traveler's cheques. I had to fork over a lot of money. Suddenly, I realized that I was nearly broke. I would very soon be in trouble financially.

CHAPTER 11

There were very attractive high-class prostitutes in the hotel lounges. I contemplated paying one or two of them for sex, but it was just a fantasy. I decided to forego it.

I met a young male university student eager to converse with an American. He took me to a café where we had a long conversation. He spoke freely about the repressive conditions in his country. The whole time, he looked around anxiously to ensure our interaction was not being observed or video or audio recorded.

I visited the Old Jewish Cemetery in the Jewish Quarter called Josefov. I saw thousands of jagged gravestones crammed into the small space of the burial ground. The cemetery was operational from the mid-fifteenth to the end of the eighteenth century. New layers of soil were constantly added to accommodate more and more graves of deceased persons who were laid to rest in caskets, one on top of the other.

On the first floor of the Pinkas Synagogue (the second oldest synagogue of Prague, which is adjacent to the entrance to the Cemetery), I saw the exhibition of drawings made in art classes by children in the Terezin Ghetto between 1942 and 1944. Most of the children who made these creative works were murdered in the gas chambers of the Nazi concentration camp of Auschwitz-Birkenau towards the end of the Second World War.

I entered the Old-New Synagogue, the oldest surviving synagogue in Europe. It is near the Old Jewish Cemetery. Legend has it that the de-animated body of a Golem was preserved in that synagogue in an attic storage room for centuries. The Golem, important in the myths of Jewish Kabbalah, is an artificial creature made of clay or mud that comes to life. It is perhaps also a forerunner of the postmodern science fictional imagination of androids, self-aware robots, and mechanical automata with consciousness. It is said that Rabbi Judah Lowe ben Bezalel of the Prague Jewish community in the late sixteenth century had special powers to animate the Golem. Its intended practical and symbolic role was to protect Jews against their enemies. One version of the Prague Golem story, as recounted by the renowned scholar of Kabbalah Gershom Scholem, goes like this:

> Rabbi Loew fashioned a golem who did all manner of work for his master during the week. But because all creatures rest on the Sabbath, Rabbi Loew turned his golem back into clay every Friday evening by taking away the name of God. Once, however, the rabbi forgot to remove the *shem*. The congregation was assembled for services in the synagogue and had already recited the ninety-second Psalm when the mighty golem ran amuck, shaking houses and threatening to destroy everything. Rabbi Loew was summoned. It was still dusk, and the Sabbath had not begun. He rushed at the raging golem and tore away the *shem*, whereupon the golem crumbled into dust.[152]

In the Book of Genesis of the Old Testament, Shem is the eldest son of Noah. Shem in Hebrew means "name." Shem has the name of Name. It is an endless recursive chain. The animation of the Golem in the Rabbi Loew legend is directly tied to language, the word, and the most sacred name. In Judaism, the first name of a person that is given at birth indicates what the purpose of their life is going to be. What was my purpose? Despite many obstacles, would I have the chance to fulfill it? Would I overcome the many adversities?

At the New Jewish Cemetery in the Zizkov district, I saw the gravestone of Franz Kafka.

Standing on the pedestrian-only stone arch of Charles Bridge over the Vltava River, I felt deeply connected with European history. I walked slowly across the bridge in melancholy meditation. The money I won in the Venice Lido casino was nearly gone. It was time to hop onto the next train: to Frankfurt, Germany. I hoped to get a cheap flight to JFK Airport in Queens, New York City, USA. Back to the New World. It was time to deal with the complex challenges of my "American way of life."

# Sociology Graduate Student in New York City

At age twenty-five, five years after completing my undergraduate degree at Cornell, I finally decided to enroll in Graduate School and pursue a Ph.D.

As mentioned, the politics department at Princeton accepted me to do my graduate studies there with full financial support. I turned down that opportunity because, based on my Jean-Paul Sartre-inspired "existentialist Marxism" ethical principles, I did not want the "free pass" to the American privileged power elite and its benefits bestowed by the Ivy League stamp of approval. I could have (and maybe should have!) studied European history of ideas with Professor Dominick LaCapra at Cornell, comparative literature at Yale, or French and Italian languages and literature at Harvard. I could have sought a humanities doctoral degree at a university in France or Italy, but, at that point in my life, I wanted to experience New York City for some years. I had the idea of studying with Jean Baudrillard in Paris, but he did not reply to a letter I wrote to him.

Literature and philosophy were the academic fields that most interested me, but I was apprehensive about them as too abstract or theoretical. I developed the notion of doing my advanced degree in sociology, which I speculated as a compromise between pure intellectual enthrallment or passion and "common sense." It would be a willful concession to practicality.

From 1981 to 1986, between the ages of twenty-five and thirty, I lived a financially precarious yet culturally enriching existence as a graduate student in New York City. It is hard to imagine surviving in the Big Apple with as little income and monetary means as I had. The financial pressures of keeping my head above water were relentless. Over five years, I slept in a dozen different abodes. Sometimes, I had a room in a shared flat in a "roommate" situation. At other times, I had my own small apartment. The monthly rent quickly increased to an amount I could not afford, so I had to move frequently. I lived in Manhattan in the West Village, the East Village, the Lower East Side, SoHo, and TriBeCa. I lived in Queens in Astoria and Sunnyside. I lived in Brooklyn in Park Slope and Sheepshead Bay.

I was an adjunct professor in the sociology department of New York University and taught undergraduate students a dozen lecture courses and seminars. Between teaching income, some fellowships and stipends, and a few part-time jobs like compiling baseball statistics, I was able to keep my marginal lifestyle going for a few years. Ultimately, I got my Master's degree but failed to get my doctoral degree. During the final phase of the Ph.D. process, I had conflicts with my professors over my dissertation's subject matter, research questions, and argumentation theses. These disagreements might have reached a workable resolution, but I ran out of money and time. There came a moment when I was flat broke. I had to drop out of the program and suspend my academic career to survive.

During that period of my life, I had an American girlfriend named Kasey for three years. She was Italian-American and from Norfolk, Virginia. After that, I had a relationship with a Puerto Rican woman named Rosalyn for six months. At the conclusion of that era, I met Helga, the German woman who became my wife for twenty-four years. I lived with her for five years in New York City. Despite her fascination with New York, Helga had no intention of settling permanently in America and planned to return to Germany. The narrative of this memoir ends in July 1991. That is when, at age thirty-five, accompanying Helga, I left America for good, finding a home in Europe, ironically in the country that killed six million Jews.

In the current chapter of the memoir, we are a decade earlier, in 1981. Having grown up on Long Island – tantalizingly close to, yet far away, from the City – and having most recently lived four years mainly in Europe, I wanted to experience New York City without being constrained by the routine of a full-time job.

Although I failed to complete the certification process that might have advanced my institutional academic career, I succeeded in the goal of participating in the life of the "City That Never Sleeps." If nothing else, I had my fervent adventures in Gotham. I lived the excitement, multicultural intensity, soaring dreams, and down-to-earth challenges of the Concrete Jungle. The soundtrack (although it's a song admittedly written and performed much later – in 2009 – by Jay-Z and Alicia Keys) is now "Empire State of Mind":

Ooooh, New York, Ooooh, New York
Grew up in a town that is famous as a place of movie scenes
Noise is always loud, there are sirens all around, and the streets are mean
If I can make it here, I can make it anywhere, that's what they say
Seeing my face in lights, or my name in marquees, found down on Broadway

Even if it ain't all it seems, I got a pocketful of dreams
Baby, I'm from New York!
Concrete jungle where dreams are made of
There's nothing you can't do, now you're in New York!
These streets will make you feel brand new
Big lights will inspire you
Hear it for New York, New York, New York! (Roc Nation, Atlantic)

As a young adult, I was excited to be (almost) a native New Yorker. Living in New York City – at that place and time – was exhilarating.

I spent most of my time in the East Village: its cafés, ethnic (Ukrainian, Indian) restaurants, music and club scenes, bookshops, boutiques, and record shops. I interacted semi-willingly with the homeless people and the population of various down-and-outers. I often went to the Mudd Club at 77 White Street in TriBeCa and CBGB's at 315 Bowery in the East Village to hear "underground" and alternative rock music. Those nightclubs were iconic countercultural venues where famous punk rock, post-punk, and new wave bands played.

Now the soundtrack is *Life During Wartime* (1979) by The Talking Heads:

The sound of gunfire, off in the distance
I'm getting used to it now
Lived in a brownstone, lived in a ghetto
I've lived all over this town

This ain't no party, this ain't no disco
This ain't no fooling around
This ain't no Mudd Club, or CBGB
I ain't got time for that now

Transmit the message to the receiver
Hope for an answer someday
I got three passports, a couple of visas
You don't even know my real name (Sire label)

Punk rock was an exciting, expressive rebellion against the money-oriented music industry. It was absurdist, anti-establishment, and anti-authoritarian. It poked fun at mainstream American culture. It celebrated the freedom of youth and the adventure of alternative – and even dangerous – lifestyles. I was not active in the center of punk rock, but it was inspirational.

I lived in the present and gave only a modest degree of thought to the future. I was passionate about my intellectual work. I wanted to be a thinker and a writer.

But I had no idea how that would connect to any "success" in the "real world" or sustainable financial situation.

During those years 1981 to 1986, I went to many dozen Mets games at Shea Stadium. I sat in the "loge reserved" section behind home plate for twelve dollars a ticket. The Mets were terrible in the early 1980s, then became "contenders" and won the World Series in 1986. My favorite players were Darryl Strawberry, Keith Hernandez, and Mookie Wilson.

> Saturday, October 25, 1986
> World Series Game 6
> Boston Red Sox at New York Mets

Vince Scully calls it on TV in one of the most famous moments of twentieth-century baseball history:

> "The winning run is at second base with two outs. Three and two to Mookie Wilson. Little roller up along first. Behind the bag! It gets through Buckner! Here comes Knight with the winning run! And the Mets win it!"

The Mets winning the 1986 World Series kept me alive.

## Engaging with Sociology

I focus here on the academic dimension of what I lived through during those years.

I was a star student in the sociology department at New York University, 269 Mercer Street, Fourth Floor, near Washington Square Park. In my first year, I was awarded their five-thousand-dollar "developmental fellowship." I was active at *La Maison Française* in the carriage house at Washington Mews, in the Department of Italian Studies, and at Richard Sennett's New York Institute for the Humanities (the latter two located on University Place, off the northeast corner of the Park).

Strangely, these felt like vocational attachments only for the present time. They did not seem to hold the promise of any future to which I could enthusiastically commit myself. French and Italian literature were excellent pastimes, but make a whole career of that? Should I become an expert on famous European novels and novelists? I thought of myself as a potential humanities scholar, but it was not straightforward how sociology was related to that. I wanted to make the connection, but the empirical social science professors were against it.

I made my way through the classic American structured Ph.D. program. Over six semesters, I took sixteen Ph.D.-level seminars and got nearly straight A's. I passed my oral comprehensive examination. I was always a good student and knew how to do those performances well.

I learned a lot from those seminars, many of which were taught by visiting professors. The courses were educational in themselves. In the end, they served no other purpose.

In the final stage of my studies, I wrote two one-hundred-page manuscripts on two different potential doctoral dissertation projects.

In my fourth year, I won the sociology department's five-thousand-dollar Ph.D. dissertation fellowship prize for my fifteen-page proposal for a thesis on the rise of the "gambling society" in America. I went on to design and distribute survey questionnaires to hundreds of respondents at Atlantic City casinos and New York City Off-Track Betting horse racing parlors. I acquired and analyzed a lot of data, deploying standard social science research techniques and statistical methods. I wrote my preliminary research report.

The two sociology professors assigned to be my supervisors did not like what I wrote. They said the manuscript contained too much "theory" and not enough about the statistics and empirical methodology of dependent and independent variables and correlations.

The famed "sociologist of power," Dennis Wrong, screamed at me in mocking New Yorker Yiddish: "Metaphor SCHMETA-PHOR!" (translation: *metaphors are bullshit*).[153] "Literary theory concepts have no place in REAL sociology!"

The famed "sociologist of medicine," Eliot Freidson, warned from his sage perch that any project attempting to bring together social theory and social science empiricism was doomed in advance.[154]

In portions of this chapter and the final chapter of the present auto-socio-biography, I will present a performative and demonstrative defense of the quality of my "sociology of gambling" work.

During my first year as a sociology Ph.D. student at New York University, I got straight A's in the following seminars:

· History of Social Theory
· Sociology of Knowledge

Professor Juan Corradi taught these courses.[155] Juan knew much about the books and social thinkers we studied but made little connection between the texts and the situation of contemporary society.

· Large Scale Organizations

Professor Wolf Heydebrand was the teacher.[156] Wolf was an incredibly lovely guy but had transformed Marxism into a dry, self-referential academic exercise with its own specialized terminology.

CHAPTER 12

· Problems of Contemporary French Society

The brilliant Claude Lefort, who was visiting from Paris, taught this seminar. In the early 1950s, Lefort wrote for Jean-Paul Sartre's journal *Les Temps Modernes*.[157] Lefort publicly debated with Sartre and broke with Sartre after the publication of Sartre's apologia for the French Communist Party, *The Communists and Peace*. As his career went on, Lefort published many important volumes of political philosophy and democratic theory. He wrote books on Niccolo Machiavelli, Aleksandr Solzhenitsyn, the philosopher of phenomenology Maurice Merleau-Ponty (Lefort's teacher), and the Solidarność anti-totalitarian revolution in Poland.[158]

The administrators, professors, and students of the NYU Maison Française had no idea who Claude Lefort was. They had no idea of the key role that he had played in French history. They had no idea he was one of the great political thinkers of the twentieth century. They had him teaching a seminar on Introduction to Modern French Civilization – a.k.a. French Civ 101. At least they didn't have him driving a taxi.

Or maybe they should have had Lefort driving a taxi (part-time). As "organic intellectuals" (see Antonio Gramsci) of the working class, we should all have

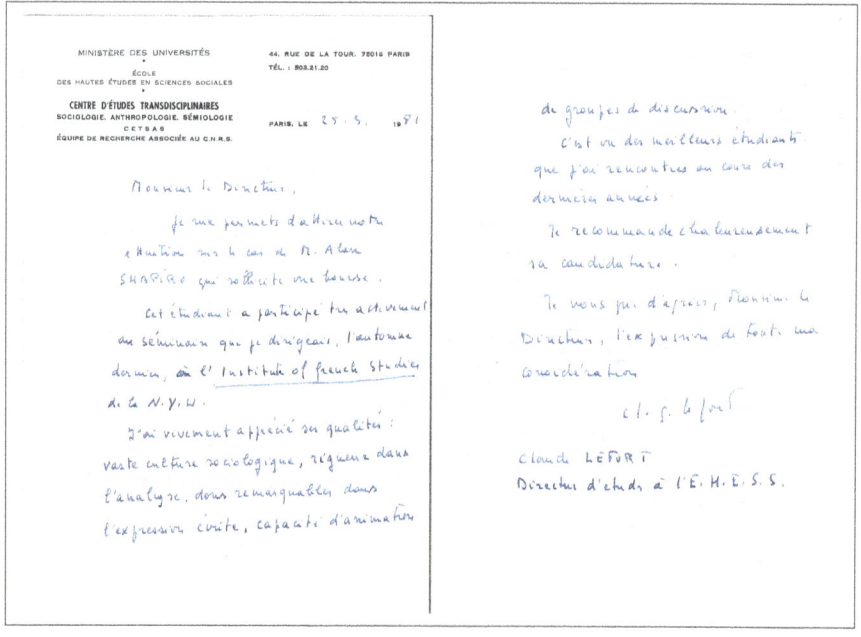

Letter of Recommendation from Claude Lefort (Alan N. Shapiro private collection)

some real proletarian experiences. After the repression of the Prague Spring in 1968, the Czech totalitarian state fired the great Marxist-humanist philosopher Karel Kosík (author of *Dialectics of the Concrete*) from his position at the university, forcing him to drive a bus to earn a living.[159] It was probably a good experience for Kosík.

The students in the class had been French majors as undergraduates. They were now in the NYU French Studies graduate program. They wanted to enter business and maybe specialize in business between America and France. They had done "junior year abroad" in France and spoke French well. Their knowledge of French history, literature, and philosophy was slightly above zero.

English translation:

> I would like to draw your attention to the case of Mr. Alan Shapiro, who is applying for a scholarship. This student participated very actively in the seminar I directed at the Institute of French Studies at NYU. I greatly appreciated his qualities: vast sociological culture, rigorous in analysis, remarkable gifts in written expression, and ability to lead discussion groups.
>
> He is one of the best students I have met in recent years. I warmly recommend his candidacy.
>
> Please accept, Mr. Director, the expression of my highest consideration.
>
> Claude Lefort
> Director of Studies at the School for Advanced Studies in the Social Sciences

Claude Lefort writes about May-June 1968 in France in his book *La Brèche*:

> For the first time, it became apparent that people could struggle without being haunted by the idea of overthrowing the power structure. It was important simply to break the partitions that isolate groups, to allow ordinarily stifled speech to circulate, and to draw up social demands. The power organs were not attacked head-on. They found themselves unhinged for a moment because those who were ordinarily submissive withdrew their alliance. The improvisation in the struggle, the freedom to act here and now without letting oneself be paralyzed by traditional ideas concerning the generalization of the struggles, the coordination of groups, the hierarchization of objectives, was also remarkable [...] Under the cover of this 'irresponsibility,' the movement was able to develop and make all of society seethe with excitement.[160]

· Society and the Literary Imagination

Victor Brombert, a professor of comparative literature (and Romance languages and literature) at Princeton University, taught this seminar at the Institute for the Humanities.[161] Like my father, Brombert fought in the Battle of the Bulge during the Second World War. He also participated in D-Day (the Normandy landings). I had read his book of literary criticism on Gustave Flaubert. Brombert was definitely an inspiring teacher.

· Research Methods I

Here is an excerpt from my report on my semester project, "Confidence in Public Institutions," that I worked on in that course:

> An intelligent foreign observer visiting the United States, initially knowing very little about American society, might suppose that there would be a relationship between people's overall confidence in public institutions and their position in the society's economic structure. He would hypothesize that such a relationship would manifest itself when the latter was measured by income, class, or any other 'objective' independent variable that gauges the distribution of wealth, privilege, and the stratified social hierarchy.
> 
> Using the data generated by the 1977 NORC [National Opinion Research Center] survey, our observer creates an aggregate dependent variable called 'Overall Confidence in Public Institutions,' which consists of a summated score compiled from people's responses to queries concerning their confidence or lack thereof in seven important separate institutions: large companies, organized religion, the educational system, the executive branch of the federal government, organized labor, the media, and banks and financial institutions. He then tests variables such as family income and social class against overall confidence, postulating that those subordinate in a society's social structure tend to become consciously alienated from that society's institutions.[162]

Two close male friends were my "cohorts" in the Ph.D. program. Chris also grew up on Long Island but was a Yankees fan. He was very interested in the critical social theory of the Frankfurt School. Chris eventually got his doctoral degree, writing a "sociology of culture" dissertation about fan communities of the rock music group Led Zeppelin.[163] Gabriel was Jewish and Belgian and had lived in Israel for several years before coming to New York. He spoke five languages (English, French, German, Hebrew, and Yiddish). Gabriel was also very interested in Jean Baudrillard's post-Marxist cultural theory of simulation and hyperreality.

During my second year as a sociology Ph.D. student at New York University, I got nearly straight A's in the following seminars:

· Social Psychology

Professor Eliot Freidson taught this course. He talked at length about the allegedly great social psychologist Erving Goffman and his seminal work in "symbolic interactionism," the book *The Presentation of Self in Everyday Life*.[164] Goffman was a master of the sociological methods of "participant observation" and "dramaturgical analysis." In daily life, we are all actors making theatrical performances to influence the impressions others have of us.

Goffman is important, but the obsessive focus on him and symbolic interactionism neglected many other crucial social psychological ideas, topics, and authors.

· Cinema and Literature

This was another great seminar taught by a visiting professor from France, Alain Robbe-Grillet. Robbe-Grillet was a filmmaker, novelist, and literary theorist. He wrote the vital essay collection *For a New Novel* and the screenplay for *Last Year at Marienbad*, a key French New Wave avant-garde film directed by Alain Resnais.[165] Along with Nathalie Sarraute, Michel Butor, and Claude Simon, Robbe-Grillet was a principal figure in the late 1950s and early 1960s literary movement of the *Nouveau Roman*.[166] The *Nouveau Roman* focused on objects in the world and was "cinematic," as opposed to the traditional emphasis of novels on internal psychology and narrative. I was impressed by the title of Robbe-Grillet's 1970 novel: *Projet pour une révolution à New York*.[167]

· Seminar in Bureaucracy

This was a smaller discussion group with Wolf Heydebrand, the sociologist of organizations. The group included passionate anarcho-Marxists (myself and others) and passionate Marxist-Leninists (the British Andrew S. and others). Wolf brought us back to Earth with his steady and dispassionate academic Max-Weberian-Marxism.

· Statistics I
· Statistics II
· Research Methods II

The sociology department was obsessively oriented towards statistical and quantitative methods. As an eternally good student, I became an expert at these, but it was meaningless.

During my third year as a sociology Ph.D. student at New York University, I got straight A's officially (according to my transcript) in the following seminars:

· Critical Problems in Social Theory
· Hegel's Phenomenology

- Field Methods
- Reading Course I

This was mainly "Independent Study." I continued my discussions with Professors Corradi, Heydebrand, and the famed Richard Sennett. I worked on the sociology of sports and gambling with Professor Edward Lehman. I had talks with Stuart Ewen, a distinguished historian of American media and consumer culture (author of *Captains of Consciousness: Advertising and the Social Roots of the Consumer Culture*), who was a fellow at the Institute for the Humanities.

The first of three areas I prepared for my Ph.D. oral comprehensive exam was "Sociological Theory." My first of three "issues" for that area was "The Controversy about Alienation in Marx." Here is the description of the "issue" that I wrote:

> Since the discovery of Marx's early writings, the various schools of twentieth-century Marxism have made different claims about the relationship between the theory of 'alienated labor' and the works of mature Marxism. In the *1844 Manuscripts*, *The German Ideology*, and the *Theses on Feuerbach*, Marx criticizes Hegel's formulation of the relation between 'objectification' and 'alienation' and points to the historical specificity of 'alienated labor' in capitalist society.[168] To what extent does Marx's conceptual framework contribute to a critical theory of society that involves a new connection to revolutionary practice?

My second "issue" for the "Sociological Theory" portion of the exam was "The Frankfurt School: A Sociology of Cultural Institutions?":

> Most recent interpretations of the Frankfurt School have tended to praise or condemn the contributions of the significant 'critical theorists.' Benjamin, Adorno, Horkheimer, Marcuse, and Habermas all sought distinctly different solutions to the problem of the relationship between philosophical concerns, cultural criticism, and political practice.[169] How can the perspective of critical theory overcome its traditional aporias and contribute to a sociology of cultural institutions?

My third "issue" in Sociological Theory was "Durkheim Between Structuralism and Functionalism":

> Emile Durkheim has been interpreted as a significant precursor of French structuralism and American functionalism.[170] A third reading of Durkheim eludes these established traditions and yields a different vision of the 'sociological imagination.' What tensions exist between sociological theory as conceived by the two standard perspectives, and what are the insights of the 'third' reading?

The second area I prepared for my Ph.D. oral comprehensive exam was the "Sociology of Knowledge." My first of three "issues" for that area was "Karl Mannheim and His Critics":[171]

What is the sociology of knowledge as conceived by Karl Mannheim? What are the most compelling criticisms of Mannheim's project? Can his thought be reinterpreted as a self-critical hermeneutics? How would this version connect to the critical 'deconstruction' of the will to knowledge and 'truth' offered by philosophers like Nietzsche, Foucault, and Derrida?[172]

My second "issue" for the "Sociology of Knowledge" portion of the exam was "The Sociology of Intellectuals and the 'New Class'":

What are the claims of the theories of intellectuals as a 'new class' formulated by authors such as Alvin Gouldner (*The Future of Intellectuals and the Rise of the New Class*), Gyorgy Konrad and Ivan Szelenyi (*The Intellectuals on the Road to Class Power*), neo-Marxists, and neo-conservatives?[173] What is the relation of the political and social practice of intellectual groups to their self-conceptions of how they articulate knowledge?

My third "issue" in the Sociology of Knowledge was "Theories of Consumer Culture":

What is the 'social imaginary' of consumer culture? How has the commodity- and media-intensive situation of late capitalism been analyzed by economists, semioticians, cultural historians, neo-Marxists, and anthropologists? How can the study of consumer culture be related to revitalizing the sociology of knowledge?

The third area I prepared for my Ph.D. oral comprehensive exam was the "Sociology of Sports and Gambling." My first of three "issues" for that area was the "Sociology of Gambling":

Most sociological studies of gambling have approached the subject in terms of deviance and pathology. They have posited a strict dichotomy between the so-called 'compulsive' gambler and the 'normal' gambler. Other writers have studied gambling in specific cultural contexts, deploying methodologies from anthropology, psychoanalysis, and semiotics. What insights have these two different approaches provided? How can gambling be understood in the context of American culture today?

My second issue for the "Sociology of Sports and Gambling" portion of the exam was "The Degradation of Sport":

Contemporary forms of organized sport have been criticized in their relation to the values and structures of the wider society. There have been studies of sport's relationship to the organization of labor, technology, social class, consumerism, commercialization, spectatorship, competition, violence, sexism, (homo)sexual repression, etc. Do these critiques present a coherent picture of the 'degradation of sport' (a term of Christopher Lasch)?[174] What notions of 'healthier' forms of play are invoked by these critical authors?

# CHAPTER 12

My third issue in the "Sociology of Sports and Gambling" was "Cross-Cultural Sociology of Sports":

> What is the place of organized mass sports in political societies today? What ideologies and discourses surround sports? What functions do sports serve in various cultural situations? What are the concepts of a potential cross-cultural sociology of sports?

## Donald Trump and Atlantic City Casino Gambling

I was especially interested in Donald Trump's involvement in the Atlantic City casino gambling industry. In 1982, Trump acquired a New Jersey casino license. In 1984, *Harrah's at Trump Plaza* opened. It was a 210 million dollar, thirty-nine-story hotel-casino. In 1985, *Trump's Castle* Hotel Casino opened. In 1986, *Harrah's* was renamed to *Trump Plaza*.

Later, in 1990, the *Trump Taj Mahal* Hotel Casino opened in Atlantic City. It became the largest casino in the world and the tallest building in New Jersey, taking up seventeen acres of land and costing one billion dollars to construct. In 1996, *Trump's World's Fair* casino opened next to the *Trump Plaza*.

Donald Trump loves McDonald's fast food and gambling casinos (Photo by Alan N. Shapiro)

The gamblers are seduced into the casino by the promise and lure of Easy Money. They are told they have a good chance to become winners. But the casino only cares about itself. Nearly 100 percent of the players end up as losers. They get *fleeced* (the process of obtaining the wool from a sheep at one shearing) and come away with nothing.

Fast money and fast food. Save on dinner so you can lose at the tables.

Since the probability of winning in the casino is small, every surprising instance of winning is highlighted and underscored through bells, lights, jackpots, video displays, computer animation, and the cheering that one sometimes hears from another roulette or craps table. When a player wins a large jackpot at a slot machine, the event is loudly proclaimed through all sorts of media everywhere in the casino. I do not lose. I am a winner. We're winners.

The casino presents a version of populist "democracy." You belong to this imaginary shared democracy – so long as you can afford the $15 minimum stake to place a wager at a table. You are allowed to sit there until your stack of chips runs out. Your right to be there, your simulated equality, is never questioned during the game. Regardless of your "net worth" outside the casino, you play against the same mathematical odds here. I can sit at the same table with the CEO of a company or a Wall Street stockbroker (although they would more likely sit at a $100 minimum per hand table). But I no longer exist as soon as I am out of chips. The casino is done with me. I can go to an ATM in the lobby and get more instant cash from/on/with my credit card. I can retire to the casino periphery of the twenty-five-cent slot machines.

The structural organization of most casino games represents a decline in sociability compared with poker. At a poker table, the player is squared off directly against his opponents. Strategy is connected with the social art of impression management. The player's gain or loss is directly linked to the outcome for one or more other players.

In most forms of casino gambling, the player is confronted either with a machine or with a highly trained representative of the casino. He or she is almost always playing against "the house." There is an atmosphere of subtle hostility among players, often an avoidance of social interaction. Disruption of the game's small rituals (hand gesturing in blackjack when to "hit" or "stick," the appropriate moment to move around one's chips) is greeted with disapproval.

In his book *Philosophy of Money*, the German social theorist Georg Simmel wrote about the attraction of money as its pure potentiality that has not been actualized, the anticipation of the future ownership of concrete things, seductive by its very abstraction.[175] Money played in casino games becomes "sign value" (units within a self-referential system) and no longer "exchange-value" (a direct stand-in for the goods and services that one could purchase with it) or "use-value" (having direct utility). The chips are narcissistic and self-absorbed, involved in a network of relations separate from the outside circulation of the economy.

Casino money is only distantly related to actual goods and services. Once it enters this local system, it can only be re-extracted into the broader circulation with incredible difficulty. Players who win tend to direct their winnings back into the game or other games or to spend them immediately. Very few will use won money to improve their economic situation. Money lost does not seem "real" either – at least while one is losing it. It feels like "play money." Players regard their losses as the price of entertainment. The mental suspension of usual associations of the value of money outside the casino is a prerequisite to participation in the game. The player must forget what she could buy or enjoy with the money invested. Even when she is winning, the player often experiences a "trance" or intoxication, which blocks her from pulling away from the tables. It is more difficult to win than the mathematical odds indicate.

Just as in gambling, where one chimerically believes oneself to be the master of one's fate, so money, in general, is regarded as a panacea and the quintessence of freedom and autonomy in American society. Money is a simulacrum (defined as an image or semblance of something that becomes more powerful and "real" than the thing it allegedly represents). One enters a self-contained and "hyper-real" system in gambling and money. The casino chips "belong" to me only in the dimmest of ways. I hold onto them temporarily, watch them gravitate away, and rid myself of my illusions of instant wealth.

Capitalism "lends" us money so we may purchase its goods and messages, follow its dictates, and live out our illusions of self-determination. Credit card transactions occur internally between the computer systems of large institutions. I monitor their transactions online in "my" accounts.

In the science fiction film *Back to the Future, Part Two* (1989), Marty McFly – played by Michael J. Fox – returns from the year 2015 to Hill Valley in 1985 to discover that all of American society has been transformed into a casino. This is symbolized by Biff Tannen's gaudy *Pleasure Paradise Casino and Hotel*. Tannen, a local bully and McFly's arch enemy, has time-traveled from 2015 to 1955 and given a copy of a Sports Almanac from the future to his younger self, thus enabling himself to win endlessly at sports gambling and become the wealthiest man in America. Think Donald Trump.

The emergence of the "gambling society" in the 1980s, in which Donald Trump was a major participant as the founding owner of three Atlantic City casinos, is related to broader social developments in American history. The abstracting and alienating forces of money, social and geographical mobility, the ideological fantasy of classlessness, uniform consumer culture and mass media, network-cable-satellite television, and organizational bureaucracy eroded social differences and local and particularistic diversity. Distinctions of status, religion, ethnicity, and community affiliation got bulldozed over.

The gambling game is the cathexis of hidden fears and anxieties. There is a concretization of psychological tension. Living on the edge is a valuable skill in today's uncertain society of economic insecurity. Casino capitalists have successfully exploited a not-so-surprising psychical and emotional reaction to a hyper-rationalized yet fearful society.

## Cultural Citizenship in Contemporary America

In my endeavor to complete my Ph.D., I wrote a second manuscript of one hundred pages about social theories of American consumerism and media culture. Since the first reports about my work on "the gambling society" were not well received by the professors (too many "metaphors" and ideas, they protested), I thought a second text could be the beginning of an alternative acceptable dissertation. I gave it the title "American *Samizdat*: From Cultural Citizenship to Cultural Critique."[176] I defined "cultural citizenship" as an extension of the historical concept of political or democratic citizenship and the sociological theory concept of "social citizenship." I wrote about the French poststructuralist critical theorists Jean Baudrillard, Roland Barthes, and Jacques Derrida. I wrote about classic works of American social and cultural theory: David Riesman's *The Lonely Crowd* (1950), Daniel Boorstin's *The Image* (1962), and Christopher Lasch's *The Culture of Narcissism* (1979).[177]

*Samizdat* is a Russian word meaning "self-publishing." In the Soviet Union and the Communist countries of Eastern Europe, *Samizdat* texts circulated as dissident or underground literature. They were often makeshift handwritten works passed from reader to reader to avoid censorship or legal punishment.

> One of my supervising New York University sociology professors read my work and exclaimed: "How dare you call your writing *Samizdat*! This is America! We are a free country! It is nonsense to call what you are writing dissident or underground! In America, you can write and publish whatever you want!"
>
> He threw a typed copy of my manuscript in my face.
>
> "And what is this French theory nonsense? The consumer society? The media culture? This has nothing to do with legitimate academic sociology! Go and read some articles in the *American Sociological Review*!"

I still think that *American Samizdat* is a great title. Herbert Marcuse's central thesis in the sociology critical theory classic *One-Dimensional Man* (1964) was that

America is fundamentally a totalitarian society, even if America is not throwing political dissidents in prison.[178] The discourses are what is totalitarian. Groupthink appears in different guises. Groupthink entails the loss of creativity and independent thinking. Psychology professor Irving Janus wrote: "Groupthink is a term of the same order as the words in the *Newspeak* vocabulary George Orwell used in his dismaying world of *1984*."[179]

The fact that I knew what *Samizdat* was showed my respect for the Russian resistance fighters.

Despite writing an award-winning dissertation proposal and two separate one-hundred-page manuscripts, my struggle to complete the final stage of getting my Ph.D. was not going well. On top of that, everyone, including the sociology professors themselves, advised me entirely against starting an academic career. The mid-1980s was a terrible time for that.

> "Maybe you can land an assistant professorship in Fargo, Oklahoma, maybe in Podunk, Iowa," one of the supposed mentors told me with smug satisfaction. "But the salary will be low, and there will be nothing interesting to do in Podunk! You will be bored out of your mind!"

The most prominent sociology professor in the area of "theory" at NYU was Richard Sennett.[180] He was genuinely an intellectual and was regarded as a leftist critic of American society. In *The Fall of Public Man* (1977), Sennett observes that the balance between public and private life has been upset in modern society. People have withdrawn into themselves. They relate to society or public life only as a formal obligation. In eighteenth-century Paris or London, according to Sennett, people interacted freely in many situations because self-realization as *sincerity* was in harmony with public life. The modern city, by contrast, is a world of strangers. Individuals are no longer capable of tapping into the creative force of the actor, the ability to play with and invest feelings in external images of the self.

I read Sennett's books with interest and attended a seminar with him. I hoped he might help me "launch" my academic career, but that did not happen.

I thought that Sennett's view about the role of technology in contemporary society was one-sidedly critical and negative. In my perspective, the blossoming social potential of new technologies places into question the socioeconomic-corporate system of "permanent jobs." The emancipatory promise of the information society is something like a true freelancer economy. Sennett's view is the opposite: he decries the rise of "the flexible man" in the hi-tech digital or hyper-modern economy.[181] For me, this was a surprisingly conservative and nostalgic stance. Unlike Sennett, I regard the skills of flexibility and continuous reinvention of the self that one acquires in a freelancer economy as a positive advance for the life quality of workers. I connect this practical suppleness that Sennett views negatively to the flexibility of consciousness recommended by Buddhism and Buddhist-inspired psychologies.

The brutal attack on the one hundred pages of cultural theory about "cultural citizenship" I wrote came from Paul Piccone. Piccone wrote a four-page, single-spaced reply. It was a full-scale onslaught against my work. He was the editor-in-chief of *Telos*, which, for a while in the 1970s and early 1980s, called itself a "journal of radical social thought."

*Telos* was also a book publisher. They translated and introduced texts of European thinkers of non-Leninist "Western Marxism" for an American audience – including two of Jean Baudrillard's books![182]

But by 1985, Piccone had, quite strangely and surprisingly, entirely shifted to the far right in his thinking. I was not yet aware of that.

The New York University sociology professor most involved with leftist, postmodernist, and poststructuralist social theory was the Argentine Juan Corradi, who had studied with the famed Herbert Marcuse at Brandeis University in the early 1960s and knew Michel Foucault in Paris.

As elaborated above, I took several seminars with Corradi. For a few years, we had dozens of lengthy, in-depth intellectual discussions in his Manhattan apartment, and while drinking cappuccino at the "NYU Institution" *Caffe Pane & Cioccolato* at the corner of Mercer Street and Waverly Place. Yet after I passed my oral comprehensive exam, Corradi took zero interest in what would happen next with my life.

Around 1982, Corradi became himself a contributor and editor at *Telos*. It was a sad turn of events for me when Juan Corradi gave Paul Piccone my paper to read and comment on rather than reading it himself and interacting directly with me about it.

Here are some excerpts from Piccone's vile attack on me:

> Your paper is an unmitigated disaster whose only redeeming feature is what comes through as the author's desperate attempt to come to grips with himself and the surrounding world [...]
>
> Do you know how brutal, merciless, and crude life was and has always been up to very recently? [...] I get the impression that the author is a product of the 1950s – groomed in suburbia in the midst of the very culture that is now attacked but which still permeates the very being of the author [...] I come from a social context [Italy] which emerged out of the Middle Ages only less than half a century ago [...] Let me clue you in: life was incredibly brutal, crude and almost unbearable [...] Any meaningful cultural critique today must not forget the achievement of Western civilization or else we risk sliding unwittingly into the realities of the Gulag or of the Khmer Rouge [...]
>
> The overlaid ideological nonsense of narratives, texts, and other recent French idiocies add nothing and make the presentation ludicrous [...] Cultural citizenship is immensely

# CHAPTER 12

```
                                    PUBLISHERS OF TELOS
        Telos Press Ltd.            A JOURNAL OF RADICAL SOCIAL THOUGHT

           P.O. BOX 3111                EDITOR, PAUL PICCONE
        ST. LOUIS, MISSOURI 63130 USA   MANAGING EDITOR, PATRICIA TUMMONS
                                        BOOK REVIEW EDITOR, PAUL BREINES
                                                        May 24, 1985

Dear Alan:

    As I promised you, as soon as I unpacked in St.Louis I picked up your paper
and carefully read it from beginning to end. That I spent so much time in
reading it and criticizing it is probably the strongest vote of confidence
that I can provide concerning my respect for your intelligence, sensitivity
and promise. But that also entails a kind of raw frankness that you may not
be altogether familiar with. So, let me welcome you in Telos. If you are
beginning to sense a set-up, you may be right: your paper is an unmitigated
disaster whose only redeeming feature is what comes through as the author's
desperate attempt to come to grips with himself and the surrounding world.
Let me quickly outline some of the main problems.

    I get the impression that the author is a product of the 1950's groomed in
suburbia in the midst of the very culture that is now attacked but which
still permeates the very being of the author. Fortunately or unfortunately,
I happen to come from a social context which emerged out of the middle ages
only less than half a century ago and where the memory of how things really
were has not yet been totally obliterated. Let me clue you in: life was
incredibly brutal, crude and almost unbearable. And if you think that things
were any better before that, then read Homer very carefully and between the
lines; analyze Achilles' life-style and compare it to living in Manhattan in
1985. Let me be more explicit: any meaningful cultural critique today must
not forget the achievement of Western civilization or else we risk sliding
unwittingly into the realities of the Gulag or of the Khmer Rouge.
```

Paul Piccone's nasty letter, May 1985 (Alan N. Shapiro private collection)

better than no citizenship at all! [...] Why do all backward societies, including Europe and Japan, not to mention the Soviet Union and Eastern Europe, desperately long for this very cultural citizenship you find so deadly?.. With all his or her shortcomings, crudities, and banalities [...] this new type of social being – the American – is much better than the ethnic groups killing each other in Ireland, Israel, and India.

My general impression of the paper is that it suffers from that unavoidable disease of over-intellectuality [...] It is the force of the fashionable, successful academic whose work is forgotten even before it is written [...] An effective intellectual can be so only if he stops being just an intellectual chewing up and regurgitating texts in Parisian cafés [...] If you were to drop entirely the sections in your paper on Barthes and Derrida, absolutely nothing would be lost [...] A page out of Hegel is worth the whole corpus of Derrida's production [...] You believe that culture and society are themselves texts – this is a pathological condition.[183]

OK, you might not like or agree with a "literary sociology" that views culture and society as "texts," but a "pathological condition"? Reading Piccone's four-page, single-spaced, heavy-handed, authoritarian, psychic assault devastated me. Approaching age thirty, I was already dealing with a multi-dimensional crisis:

(1) No emotional or financial support from my parents for what I was doing with my life
(2) No money to pay my rent or buy essentials
(3) Emotional turmoil with two girlfriends
(4) The narrow-mindedness of the American sociologists who had no respect for my more literary, media-theoretical, and "French" version of sociology
(5) The right-wing political climate of the Reagan era
(6) The lack of social science professorship openings in America in the 1980s

His letter was the straw that broke the camel's back. After reading it, I decided to stop work on my two dissertation projects and give up the quixotic quest for the Ph.D. I was done.

Here are the opening four paragraphs of my manuscript on "cultural citizenship" against which Piccone launched his attack:

> The damaged life of the fragmented individual, performing his duties in the system of production, is circumscribed by a universe of self-referential images and solipsistic spaces that sustain his socialization in that system. Walking amidst the skyscrapers and luxury boutiques of the Upper East Side of Manhattan, one has the impression of being in an underground city of loyal citizens, cut off from all other possible space and stories, real in history or imagined in dreams. Dwarfed by the immense monoliths, beset from all sides by an unblemished futuristic-technologized décor, the individual is called upon to exonerate his presence in this complete world. He is compelled – within himself, concerning others, with respect to organizations, and to the undivided ambiance – to make known which qualities and skills give him the right to participate in the scramble to belong.
>
> From how he speaks and dresses to the technical knowledge and credentials that he carries, he is like an African American in the 1950s being scrutinized by White Cops, a Kafka man permanently On Trial, seeking to decipher the intricate workings of the Court. Immersed in a world of unfathomable complexity presenting itself as a closed perfection, the aspiring Cultural Citizen of contemporary American society must conceal his human weaknesses, scratch and claw to plant a stake, and strive to resemble in his being the rigid geometry of this system of survival [...]
>
> The processes and modes of legitimation, participation, identity, and solidarity in late capitalism cry out for thorough re-examination in the context of a renewed Critical Theory of American Society and Consumer Culture. This project involves revitalizing and rethinking the vanishing genre of cultural theory and elaborating a concept of cultural citizenship. Since the American and French Revolutions at the end of the eighteenth century, political citizenship has dominated our thinking about social participation to the virtual exclusion of other approaches [...]
>
> But the New Left in the 1960s, at least in its most lucid and audacious moments, initiated a critique of politics that continues to ring true. Those who partook in the assemblies and action committees in France in May-June 1968 voiced the demands

for 'participatory democracy' of the early Students for a Democratic Society (SDS) in America, marched in the Civil Rights Movement or against the Vietnam War, or actively refused the collaboration of the major political parties during the 1977 student uprising in Italy, shared a critical perception of politics as having become a separate sphere, a Simulation of Democracy, a realm reserved for experts and professionals and divorced from everyday life.[184]

## The Sense of an Ending

In New York City in the 1980s, everything was getting more expensive. It was no longer possible to find a cheap apartment. The deli coffee for fifty cents was superseded by the upscale yuppie cappuccino for three dollars. The income I had from teaching assistantships was not much to begin with. Now, it was over. You were allowed to teach for three years. They then gave your position to a second-year graduate student. The few thousand dollars I had from the dissertation fellowship were gone. I used them for living expenses. I had taken out ten thousand in a student loan but was wary of borrowing more and putting myself in debt for life. I had a few part-time jobs, but they barely helped. I drove a car for a wealthy disabled woman. I ghostwrote and compiled three books on the history of baseball statistics (the statistical histories of the New York Yankees, the Brooklyn-Los Angeles Dodgers, and the St. Louis Cardinals) for Gene Schoor, a sportswriter of biographies of famous athletes.[185] Making ends meet was becoming unworkable. I tried for a while to postpone the obvious inevitable: time to quit academia, quit the Ph.D., go to midtown Manhattan and get a full-time job with a salary.

One summer evening, I went with Gabriel to the Meadowlands harness racetrack in East Rutherford, New Jersey. I won $700 on a "triple." You must correctly pick the top three finishing horses in a race in exact order. That put some cash in my wallet. Fortunately, my big win was the last race of the night's program, so I had no negative opportunity to lose the money back immediately. The next day, I went to my landlord's office in Park Slope and paid one month's rent in cash.

One year into my graduate studies, I met Kasey at a party and started a romantic relationship. She was in love with me and would have stayed with me long-term, but I broke up after two years and nine months. I liked her a lot but was not ready to make a commitment. I definitely mistreated her by staying with her for as long as I did, even though I knew that it was not going to last. I was lonely when I first met her, and having her in my life provided me with company and security. I admit that it was not moral behavior on my part.

Kasey had a full-time job as a typesetter. It was a profession that used to require the pre-digital skills of arranging blocks of wood or metal onto plates. She was left-liberal politically, a warm and understanding person, and very much into punk rock music. She listened to vinyl records of Elvis Costello, the Ramones, and the Sex Pistols. I converted Kasey to fandom of *Star Trek* and baseball. We went to many night games at Shea Stadium. She had two cats and lived in a tiny apartment above an Indian restaurant on Sixth Street in the East Village. The cats were free to come and go via an open window. There was always the fear that the landlord would evict the tenants from her building and convert all the units to co-op apartments.

Rosalyn was a fellow Ph.D. student in the NYU sociology department. She was a radical leftist, deeply intellectual, and very well-read. She criticized the conservatism and faux liberalism of the professors and the department's bureaucratic academic mentality. We started a passionate, sensual relationship that lasted six months. We had meaningful and intense discussions about life, philosophy, America's social problems, childhood and family traumas, and writers and artists who profoundly affected us. We had many genuine moments of feeling deeply in love.

Subsequent events, however, tore me apart emotionally and psychologically as the story unfolded. Before our involvement, Rosalyn had been with a British boyfriend named Andrew, who was also a Ph.D. student in our department. I knew Andrew from a seminar we were in together. When Rosalyn and I first started dating, she told me she had broken up with Andrew a few weeks before. It soon became apparent that Andrew did not believe or accept that they had broken up. During the months that followed, Andrew "stalked" us. When we returned to Rosalyn's apartment after an evening out, he was sometimes waiting in the lobby to harass us. At other times, we were in the apartment, and he would ring her doorbell incessantly.

> One time, I encountered Andrew early in the evening on a street near Washington Square Park. I tried to be aggressive and "stand up" to him, forcefully telling him to stop his stalking, even mildly threatening him with physical violence.
>
> Andrew laughed at me. "She's *my* girlfriend!" he said. "It is destiny that she and I will be together. You are the NEWCOMER!"

After several months of amorousness and harmony, Rosalyn and I had our first serious argument. Then she announced that she was "going back to Andrew." She left my Park Slope apartment in a huff and went and stayed with him for forty-eight hours until they fought. She returned to me in tears and asked for forgiveness.

"You are the man of my life! You are the one I am truly in love with!"

I said OK, and I took her back. But if she ever did it again, that would be the end for us.

Several weeks later, the pattern repeated itself. We argued about how, in her view, I allowed my mother to speak nastily to me in an unpleasant phone conversation that put me in a bad mood. Rosalyn "went back to Andrew" for four days. She came back weeping and in distress. She apologized again and begged for a fresh start. I did not take her back. It was over for me.

I drove Rosalyn in my small, white Toyota Corolla to Newark Airport in the middle of the summer of 1984. She was going to visit her family in Puerto Rico. We said goodbye at the gate. It was all too painful. I decided I would never speak with her again.

I had become very attached to her in body and soul. Now it was finished. I suffered physical withdrawal symptoms like skin rashes. I was heartbroken.

I was always struck by the similarity between *Rosalyn* (Puerto Rican girlfriend) and *Roslyn* (hometown).

Decades later, I did indeed speak with Rosalyn again. We made contact on one of the "social media" platforms and developed a "pen pal" friendship. Rosalyn is, in many ways, a fine person. She apologized for how she treated me in 1984, and we have achieved forgiveness.

Rosalyn is a valuable friend, among other reasons, because she is a living witness to what I experienced – and what we both experienced – with the 1980s NYU Sociology Department and its professors. Rosalyn lived through things very similar to what I went through and shares my critical view of that institution. Her solidarity means a lot to me.

Although I self-critically take responsibility for events in my life, I think it is fair to call what we experienced *mistreatment* by the NYU sociology department professors.

Eighteen months later, in January 1986, I met Helga for the first time while on a short trip to Germany. I decided spontaneously to fly to Frankfurt for a two-week visit. Professor Juan Corradi from the sociology department was coincidentally on the same flight with me. He was supposed to be my dissertation advisor, but we were no longer meeting for discussions over cappuccino. We chatted briefly before the plane's departure. I asked Corradi why he had given my written work to Paul Piccone, who ended up "ripping it to shreds." Corradi had no reply.

I first encountered Helga at a party at four o'clock in the morning. We talked and danced for a while. We met the following morning for a late breakfast in the Rotlint Café in the Nordend neighborhood of Frankfurt. In later years, while recalling that scene, she always made fun of me for having absent-mindedly "massacred a croissant" with a butter knife while I focused obsessively on our conversation and ignored the food in front of me.

My connection to Frankfurt began in 1980 when I stayed a few months in the apartment of my friend Bernie at 96 Perry Street in the West Village in Manhattan. Bernie stayed overnight most of the time with his life partner on the Upper West Side, so his Village apartment was empty, and I stayed there alone.

One time, Bernie and I spent the day showing Daniel Cohn-Bendit and his three German traveling companions around New York City. Cohn-Bendit was the Jewish German-French charismatic leader of the student movement of 1968 in Paris who later became an important politician in the Green parties of both large Western European countries.[186] The contact with Cohn-Bendit came about via a close friend of Bernie's in Detroit, a professor of labor economics who had ties to international left-socialist circles. "Dany le Rouge," as he was dubbed by the media in 1968, spoke very little English. He talked all afternoon with his friends in German, but I tried to engage in discussion with him in French. He ignored my attempts to converse, which disappointed and bruised my ego.

Other interesting and congenial visitors from Frankfurt came to Bernie's apartment. Siegfried stayed for one week. We became friends. A professor of the history of the German labor movement slept in the flat. Later, when I had my apartment in Park Slope, Brooklyn, Klaus and Conny, who were friends of Siggy, visited and stayed with me for several weeks. Josef S., an artistic painter from Frankfurt, was my guest for one week, and we became close.

Helga's passion was literature. She wanted to build a career as a translator of literary novels from American English to German. She planned to live in the USA for five years to achieve a nearly complete knowledge of English. After our initial exchange in Frankfurt, we wrote letters for a few months. Helga came to New York, and we began our romantic involvement. Helga was a great person and companion. She was idealistic and had dreams in life, but was also very reasonable and practical. She had good moral values. She was sophisticated, political, cultured, and well-read. We harmonized very well and seldom argued.

Helga and I lived in an apartment in Sunnyside, Queens for four years. During the second half of my decade living in New York City in the 1980s, I had two full-time jobs in midtown Manhattan. Sunnyside's location was convenient for getting to midtown in fifteen minutes – assuming that the train did not break down or get delayed due to mechanical problems. Sunnyside was also advantageous for traveling to Mets games in Flushing, Queens.

We married mainly because Helga needed a "green card" to work in the United States. We had our wedding ceremony at the Ethical Culture Society in Brooklyn Heights, a venue I chose because that organization seemed to match my Jewish, atheist, and secular moral-humanist beliefs. Helga agreed with the decision.

Helga worked part-time at a German financial investment company on Wall Street. She studied and completed a Master's in translation studies at the City University of New York Graduate Center.

Helga and I stayed together for more than twenty years. I was extremely fortunate to meet such a wonderful woman. I can honestly say that she saved my life. Ultimately, I went with her to Germany in 1991 at age thirty-five when this memoir ends. Helga gave me love and stability. She got me to a livable situation and the chance to carve out a good life in Europe. This was a crucial achievement in my life. It consummated my long-sought "trap door" and "escape hatch" way out of America.

Educationally and vocationally, I had the ignominious status of A.B.D. (All But Dissertation) but was proud of it. It's a complex condition, but considering how things are in the academic system, you can wear it as a badge of honor.

My failure to get my NYU sociology Ph.D. and dropping out of academia were blessings in disguise. In the following years, I had full-time jobs in Information Technology in the (tempestuous) business world. I became a software developer. This meant I learned about technology from the inside. I learned about digital, virtual, cybernetic, and informatics technologies. Academic professors in cultural studies and media theory write about technology while observing it from the outside. Ensconced in the ivory tower, they don't really know the object of their investigation. If I had continued on the "one-way street," I would not have known technology. Later in life, when I returned to composing the books I wanted to write, I became a better philosopher and social thinker than I would have been. I am grateful for the circumstances that led me to quit academia at age thirty.

# Greenwood Mills Marketing Company

I was thirty years old. I failed at everything consequential that was either set up by my socio-economic circumstances or which I tried while striking out (in both the baseball meaning of that phrase and the meaning of starting out on a new or independent course of action) on my own. Some of the failures were due to my rejecting chances that I did not feel to be ethically or existentially correct. Many of the possible paths filled me with dread when I contemplated them. I had no stomach for them.

I intentionally passed up on many of the "golden opportunities" available to me. I could have studied civil engineering and gone to work with my father, and I would have become a multi-millionaire. Having been accepted at MIT at age fifteen, I could have studied another engineering discipline or a natural science field and started a lucrative technoscience career.

In another register, I could have gone for my Ph.D. in European Intellectual History at Cornell University with Prof. Dominick LaCapra, whom I admired greatly and with whom I studied as an undergraduate. But I sensed that LaCapra was too much of a father figure or a person of "psychoanalytic transference" for it to be a non-problematic relationship.

After the money from the winning harness racing bet at the Meadowlands ran out, I was flat broke again and could not pay the rent on my run-down tiny studio apartment in Park Slope, Brooklyn. I lived on the top floor of a non-renovated brownstone walkup at 127 St. John's Place between Seventh and Eighth Avenues. There were scattered large holes in the building's roof and in my ceiling. When it rained hard or there was a snowstorm in winter, I had to set up buckets on the floor to catch the water. The landlord failed again and again to make repairs.

The time had come. I had no choice but to go to midtown Manhattan and find a job. The decision to take that decisive step left me permanently branded with the worthless and unenviable title of an academically disgraced A.B.D. – the acronym for "All but Dissertation."

I bought the Sunday *New York Times* with the fat Classified Section for Help Wanted Ads. I grabbed a coffee and a Cheese Danish at a breakfast counter and perused the job listings. Sitting on my swivel chair in the Brooklyn diner, I could find nearly zero work positions for which I was qualified and that I could accept in my mind. I did not want to be a secretary. I could be a collections assistant at the Museum of Natural History on the Upper West Side for an annual salary of $16,000. Any college degree in the humanities would get an interview there. But that low-income level would be coming too far down in the world.

But wait – here was an ad from a company looking for someone who knew French and Italian! The bonus was that no employment agency ("middleman") was involved. I resolved to call them early Monday morning.

It was the Greenwood Mills Marketing Company. Greenwood Mills, Inc. is a privately or family-owned textile company based in Greenwood, South Carolina. The company was founded in 1889 by William Lowndes Durst as the Greenwood Cotton Mill. Throughout its history, it has always been a flourishing textile business. Greenwood is proud of the continuity and low turnover of its workforce. Many employees have worked for the company for decades. Many belong to families with three or four generations of what the company calls "associates."

A team of several dozen salespersons in the New York office marketed the fabrics. The Marketing Company occupied three floors in the high-rise office building of 111 West 40th Street, between Sixth and Seventh Avenues. About seventy-five people worked in the Manhattan office.

I dressed in a jacket, white shirt, and tie and took the subway to midtown. I successfully completed the interview with Edward and Norm, and they offered me the job.

They both impressed me as very nasty men – Norm even more. But I decided to tough it out. I desperately needed the job. I saw no other way forward with my life at that juncture. I needed a steady biweekly paycheck. I had to get at least one year of so-called "experience" on my resumé to look for a better, less stressful, or higher-paying position. A position where I would not be subjected to assholes bossing me around. A position that enabled me, at last, to pay my rent and have some discretionary funds left over to buy myself maybe a large bucket of Kentucky Fried Chicken or a week at Disney World.

"The best way out is always through," as the poet Robert Frost said in a famous quote.

I was honest with Edward and Norm about my biography. The choices I made led me to this "shipwrecked" situation. I regretted that I had not prepared myself with any forethought for any career. I regretted that I stayed with sociology so long. It ejected me and left me nowhere.

Edward (Harvard Law School Class of 1963) was an attorney and the company's executive legal counsel. Norm was the chief accountant and controller. Together, they were the bosses of all administrative and operational areas.

Edward and Norm seemed to delight in my misfortune and predicament. They knew that I needed the job badly, and they would enjoy lording it over me. They started me on $20,000 per year. That was very little income. I would not be able to survive on that salary. But I had no choice. My back was against the wall. It was my time to swallow shit.

Denim was a hot-selling fabric. The company was exporting denim to buyers in France and Italy. My job was to talk on the phone in French and Italian to those European customers.

At the end of the first two weeks, Edward called me into his large corner office. He told me the company had a more pressing need and a more important job for me to do. AT&T had just installed a System 75 Private Branch Exchange (PBX) on the premises. The company henceforth had to manage its own telecommunications. Edward needed someone to figure out and then administer the system. He handed me the three-ring binder Implementation Manual. My new job title was Communications Manager. I was proud to have a managerial title. He also put me in charge of many other areas, from Personal Computers to rodent control. He promised me a substantial salary raise in six months.

Many American companies became responsible for configuring their own telephony in the 1980s following the 1982 court decision that broke up the Bell System monopoly. The United States vs. AT&T antitrust lawsuit instructed the parent company to divest itself of the seven regional "Baby Bells," which became independent and continued to provide local telephone service. AT&T, operating nationally, provided long-distance call service.

I spent much time in a metal-caged closet in a poorly lit backroom. It reminded me of the unfinished basement of my childhood, the back stairwell at the synagogue where I hid out, the library stacks at Cornell, and the good old days of sorting hangers in the third sub-basement of Macy's. But it was different this time. I was learning "real-world" skills for advancement in capitalism. The skills were a hybrid of theory and practice.

In my clandestine lair were steel cabinets and a computer console, entanglements of wires and cables, microelectronic switching cards, and modular power units. There was a standalone backup power supply. I tinkered with the PBX's internal switching network, adding features for individual users or sales groups at the request of management or the users.

I configured the Attendant Console used by the receptionists and incoming call operators. I set up a multitude of system features, functions, and services: alternate call routing, automatic callback, call forwarding, call waiting duration, night service options, personalized ringing, recorded announcements, voice message recording and retrieval, remote access to messages from outside the office, random uniform departmental call distribution, and the Personal Computer (PC) / Private Branch Exchange (PBX) Connection. I filled out system forms on the system terminal screen: the abbreviated dialing lists, allowed call lists, authorization codes, call coverage answer groups, and modem pool groups.

I roamed the company's three floors, crawling under desks and connecting telephones via modular cable to the system's nodes.

My work week did not begin at 9 AM on Monday. It began at midnight on Sunday evening. The Data Processing Department in South Carolina started a new practice of sending all updated information on textiles produced in the last days, which were available to be sold during the forthcoming week, to an IBM line printer in the New York office. I had to monitor the behavior of that clunky printer, keep it supplied with paper, and resolve paper jams. The fabric types and quantities data were engraved onto "green bar" computer paper with alternating green and white horizontal stripes and tractor feed holes at the edges. It took several hours for all the wide-format reports to print. There were multiple pages designated for each salesperson.

Alone in the wee hours of the morning, I went around to all the desks on the vast sales floor, one level below my office, distributing each printout to its assigned recipient so he or she would be ready to start selling to the max at 9 AM sharp.

I had to continue working my regular shift until 5 PM on Monday. I had to be back at the Greenwood Mills Marketing Company on 40th Street at 9 AM every Tuesday to Friday.

> One time, I dawdled while reading the newspaper, or the subway train from Brooklyn got delayed, and I walked into the office at 9:15 AM. Edward stood there waiting for me and made an exaggerated gesture, curling his arm over the imaginary ghost of my shoulder and looking at his wristwatch.
> "You're late! That is unacceptable!" he exclaimed.
> "I am sorry, Sir! It will never happen again!" I replied.

Faxing was a crucial communications technology in the 1980s. As the Communications Manager, I supervised three fax operators who had that low-paying job because they were mere high school graduates and, sadly and *de facto*, because they were African Americans. Yes, racism was institutionally built into the system. Salespersons came in and out of the fax room constantly, bringing their documents to be scanned and transmitted – or picking up their incoming paper copies of the bitmap image reconstructed from the audio-frequency tones relayed via telephone from sending to receiving machine.

## Enter the Personal Computer

Almost everyone in the company, from the fabrics marketers to the fashion designers and artists on the upper floor to the secretaries, was in the process of getting an IBM Personal Computer – or a less expensive "clone" from Taiwan or Dell or Compaq Computer of the coveted newfangled IBM device. Edward assigned me the task of learning everything I could about how PCs worked. I was responsible for unpacking them from boxes, setting them up physically, installing software, troubleshooting problems, replacing defective parts, and explaining to employees how to do their work on the computer. I learned how to insert hardware components such as internal modems, fax cards, and extra Random Access Memory into the expansion slots on the motherboard. I learned the PC-DOS or MS-DOS operating system commands. I learned my way around WordPerfect word processing software and Lotus 1-2-3 electronic spreadsheets. I learned how to maintain a relational database management system.

The IBM XT Personal Computer, with its 8-bit "data bus" and 16-bit 8088 microprocessor, was based on the 8086 chip of the original IBM PC introduced in 1981. The data bus is a subsystem that handles information exchange between components like the CPU and working memory. "Bits" are the binary digits with a value of 0 or 1, the smallest data representation units in digital computing.

The IBM AT Personal Computer, introduced in 1984, was designed around the Intel 80286 microprocessor. The AT had a 16-bit bus. 8-bit or 16-bit meant devices architected around integers, registers, or memory addresses that are eight or sixteen bits wide, respectively.

There was the next series of PCs based on the 32-bit 80386 Intel microprocessor.

Some PCs had a hard drive. Some had only two "floppy" or external disk drives.

In the 1960s, there were semiconductors, mainframes, and minicomputers. The 1980s saw the introduction of the Personal Computer to the marketplace. It was a cultural revolution. The PC was advertised and sold to the public as a tool for personal empowerment, visual design, and creative expression. The computer was transformed from a calculation machine to a device for media consumerism and individual daily life self-administration.

The 1960s was the era of the classical mass media. With movies, advertising, and a TV system of very few channels, the ordinary person was trapped in a mediated relationship with the dominating "spectacle" (Guy Debord). Citizens of media culture were in a fundamental situation of spectatorship and passivity with respect to the power of the screen and the endless alleged "cornucopia" of consumer objects. In the 1980s, the consumer was encouraged to participate fully in the spectacle of cultural-economic activity as a media producer. The spectators were tasked with producing and distributing the images of the "integrated spectacle."

In his 1987 book *The Ecstasy of Communication*, Jean Baudrillard characterizes the era of digital media and online technologies as an interactive performance where the individual stationed at his computer becomes a self-managing and self-surveilling node of a relay switching network, micro-managing his own little world of operations and desires.[187]

The Graphical User Interface (GUI) of Windows and the Mac – with its mouse and touchscreen input, drag-and-drop, desktop metaphor, software applications, hypertext, and information presented as the multimedia juxtaposition of text and image – replaced the text-based command-line interface.

During a timeout in the third quarter of the January 22, 1984, Super Bowl (the championship game of NFL professional American football and a mega-spectacle watched on TV by tens of millions of people), a 60-second science fiction-themed commercial for the Apple Macintosh Personal Computer was shown. "1984," as the TV spot was called, was directed by Ridley Scott, the famed director of *Blade Runner* and the *Alien* movie series. The mini-film made reference to Fritz Lang's 1927 pioneering science fiction film *Metropolis* and George Orwell's 1949 dystopian SF novel depicting a totalitarian society ruled by the ubiquitous menacing image of "Big Brother."

In the middle of a drab, grey industrial-era society of depressed and mindless-looking drones who march in step surrounded by surveillance telescreens or sit gazing at and listening passively to the demagogic oratory of Big Brother, a heroine in full color and athletic clothing, chased by armed state police wearing intimidating black uniforms, runs onto the scene. She tosses a sledgehammer at the big screen, causing an explosion of whitish-blue light and smoke, symbolically representing the revolutionary overthrow of the repressive order of things.

A scrolling text appears on the Super Bowl viewer's TV screen, accompanied by a voiceover reciting the same words:

"On January 24th, Apple Computer will introduce Macintosh. And you'll see why 1984 won't be like *1984*."

The rainbow Apple logo appears as the commercial ends.

The totalitarian society portrayed in the Ridley Scott Mac commercial is meant as a semi-serious metaphor for how things were when IBM was the near-monopoly and archetypal company of the computer industry. Computers were widely imagined through the image of impersonal bureaucratic rows of mainframes,

GREENWOOD MILLS MARKETING COMPANY

Still image from the January 1984 Super Bowl TV commercial "1984," introducing the Apple Macintosh Personal Computer, director Ridley Scott (Apple Computer, Inc. Academic Fair Use)

set up in the service of Big Business and Big Science, deployed for scientific calculations, accounting, and "number crunching."

The Personal Computer was about to change all of that.

I became an expert in writing code in a "scripting language" for file transfers between remotely located computers using the BLAST software and communications protocol. BLAST worked with the asynchronous RS-232 comm ports of PCs and dial-up modems that were popular in the 1980s. BLAST had many advanced features to guard against errors caused by "noisy" telephone lines. It was known for its bit-oriented data encoding and retransmission of corrupted data blocks. File transfers could execute in both directions simultaneously.

## Not the King of the Roost

My boss, Edward, kept loading more responsibilities onto my plate. He put me in charge of the rodent and insect problem in the three-story office space. I interfaced with the rat and cockroach exterminators. Edward also put me in charge of miscellaneous purchasing. Anyone who wanted more pens, pencils, or a coffee maker had to come to me and justify their request.

## CHAPTER 13

The company sent me on a trip to South Carolina to see the setup of the production mills and the Data Processing Department. I had fantastic shrimp and grits for breakfast at the hotel and an excellent beef bourguignon for dinner.

After six months, Edward called me into his office for the salary review. He congratulated me on doing excellent work and announced I would get a 33 percent raise to $27,000 annually. Getting such a big percentage raise was highly unusual, and he needed special permission from the company president. I felt I had accomplished something, but it was still very little money.

I worked at Greenwood Mills Marketing Company for eighteen months. I had to leave to escape the authoritarianism of Norm the company controller. I could only stand his bullying of me up to a point. It was making me physically ill. Norm was capable of being nice and respectful, but not to me. I observed him being friendly and civil to others on the same level or above him in the world's hierarchy. But to those below him in the pecking order of the urban jungle (like me), he was a nasty son of a bitch. In this respect, he reminded me of my mother. She could be charming when that behavior helped her get something she wanted from someone.

> Norm was seventy years old but still charged up with boss energy. He commuted in the subway from Brooklyn to midtown Manhattan and started his workday every morning at 6:15 AM. "If you want to get ahead in this world, Alan," he told me, "you have to be willing to give up some of your personal time." "You're gonna learn, Alan, you're gonna learn," his forefinger wagged at me.
> 
> "You're not an adjunct professor at NYU anymore, Alan. I know what that's like. I taught accounting. As a teacher, you're *king of the roost*." He placed great emphasis on those last four words.

Often, I would be carrying out my duties as a manager or technical expert – configuring the telephone system, setting up a new Personal Computer for an employee, supervising the fax operators, or distributing the fabric availability printouts – and Norm, wandering into the physical space of whatever I was doing, would invoke and assert his authority over me by taking over the task, pushing me aside, condescendingly giving me orders, in front of others.

Norm's nasty bullying personality must have developed early in life because he was short. In our culture, boys who are short of stature are mistreated. They often grow a hard-shelled personality as the armor of a generalized defense mechanism. I was inclined to psychoanalyze Norm in this way since I was very short in adolescence. Once again, I felt gratitude to the universe for having grown seven inches during the summer at age sixteen. That spared me from sprouting some personality deformation to protect myself in the cruel world.

I made one friend in the company. Paul and I often went to lunch together. He was my age, about thirty. His life and financial situation were equally shipwrecked like mine. He got his undergraduate degree at Harvard, studied philosophy, and never made any concrete plan for his future. He had not gone to any Graduate School. He was stuck in a tedious job that he hated. He was in the Credit Department. Paul spent his workdays on the telephone with potential wholesale buyers of Greenwood Mills fabrics, determining the degree of their creditworthiness.

> I've got a plan, Alan, to get out of this corporate nightmare world," he told me. "I'm going to go to Law School at night." Then I'll become a lawyer and HANG OUT MY SHINGLE." He elaborated on the concept of "the shingle": "I'd rather be self-employed with little income than in this stressful rat race with a big salary. What's the good of money if you are miserably depressed all the time? Twenty-five thousand a year will be enough.

Paul's love life was also a disaster. He lived with a childlike woman who vexed him and who was uninterested in sex. But he felt morally responsible for her welfare and could not leave her. He aspired to meet a second woman for whom he felt more passionate, and they would live together in a household of three. It sounded like a crazy fantasy that would never happen.

On the way back to the office after lunch, Paul often exclaimed: "Back to the snake pit!"

I plotted my departure from my first serious full-time job in the business world. I figured I could get myself into a less stressful situation AND for more money. I started to teach myself computer programming. I started to go to interviews.

I could not stand that job for one more minute, yet the existential experience was valuable. I learned that staying alive meant a lot to me. I was willing to walk through hell for a period of time if it was a step toward my survival.

Away from the university's Ivory Tower, I experienced the factuality and harshness of the "real world."

I was starting to learn about technology for real. I was gaining familiarity and knowledge I would have never had in the academic environment. This was a priceless advantage for becoming later a thinker of technology and society with genuine insights.

CHAPTER 14

# Wall Street Computer Programmer

Before I could quit Greenwood Mills Marketing Company, I needed to get a new job offer. That's how things worked in the capitalist New York City business world of "hire and fire." I had no money saved and could not afford any time without a paycheck. While continuing to do my eight-hour shift per weekday plus the supplemental Sunday nights, I went to interviews for positions at other companies on the side. Sometimes, you went to an employment agency or "headhunter" who (after a filtering process) sent you to an interview with the "principal." Sometimes, a phone call to the number given in the *New York Times* job ad puts you directly in contact with the company offering the job. Compared to eighteen months before, I was now knowledgeable practically about technology (PCs, telecommunications, scripting languages), so there were many job announcements in the "newspaper of record" to which I could respond.

I got a job offer from a small company that provided technical support for phone (voice and data) systems. I accepted their offer. I quit the Greenwood Mills job, giving my boss the famous "two weeks' notice." Before starting the new job, I got a second offer from a second company I had interviewed with. It was a software company with a medical office management application that it sold to physicians in private practice. I had to make an agonizing decision between the two simultaneous offers. I decided to take the second one (the one in the medical field). The salaries were about the same. Both offered five thousand dollars a year more than my Greenwood Mills salary. The second job seemed somewhat more interesting content-wise.

I had to tell the owner of the company whose job offer I had already accepted that I had received a second offer and, after agonizing over it, had decided to go with the other job. I was in midtown Manhattan on Forty-Second Street on my lunch break and called the boss from a pay phone on a street corner. He got hopping mad. He claimed I was behaving unethically by turning down his job after having already accepted it. He cursed me out prodigiously. He accused me of all kinds of evil things. I tried to defend myself. I said:

> This is the business world. This is standard behavior. This is what people do when job hunting. Taking a job I don't want because of some supposed ethical principle doesn't make sense. This is the only right I have as a worker in this economic system: the right of the worker to quit a job at will. Like everyone else, I must do what is best for my career.

There was no calming him down. He kept screaming vulgarities at me. I reluctantly hung up the phone while he kept cursing.

The new job (the one of the two that I took) was a disaster. It was a small startup company with fifteen employees. They packaged their medical administration software with the "dedicated" hardware of a Personal Computer they sold to the clients. Marking up the price of the PC significantly increased their profit margin. The bosses expected me to do a lot of physical labor (carrying boxes, assembling industrial metal shelving) that they had not mentioned at the hiring interview. They wanted me to work from 8 AM to 8 PM six days a week. When I explained that I was neither physically nor mentally capable of keeping up with such a schedule, they fired me. The job lasted one week.

I had no job, income, or money in the bank. I didn't think at all about returning to academia to resume work on my sociology doctoral dissertation. I knew that I would get another job in the business world. I had eighteen months of experience on my resumé. I had the technology knowledge. I had already gotten two job offers (although neither position had "worked out"). Nonetheless, it was a tricky and precarious situation because I was broke. I did temp work five days a week for two months. Temp agencies sent me on several assignments.

At one point, I was desperate for income and had no work. I answered an ad for a temp position to do word processing on a Mac at a large law firm. I had zero experience with Mac computers but lied and said I did and knew them well. My appointment at the temp agency was at 4 PM. They were going to give me a Mac test. It was 10 AM. I raced to a bookstore and bought a book about navigating the Mac operating system user interface and mapping keystroke combinations. I ran back to my apartment and crammed the information in the book into my head. I sprinted to the temp agency appointment, passed the Mac test, and got a two-week gig that paid a decent hourly rate. I performed those two weeks successfully.

I continued to go to interviews for "permanent" jobs.

I was interviewed by Larry, the Director of Data Processing at Neuberger & Berman, a Wall Street stock brokerage firm specializing in portfolio management for super-wealthy clients. That interview was a blessing and "saved my life." Larry was amiable and empathetic. I was honest with him about my biography and "the mistakes I had made." He saw on my resumé that I was accepted to MIT at age fifteen and studied there for two years. That convinced him of my innate intelligence and aptitude for mathematics and logical thinking. He believed that I had potential as a computing specialist who could add value to the department. He was himself an MIT graduate. Larry hired me. He would take me on as a software programmer, even though I had no programming knowledge or experience. I could self-teach while working. I could "learn by doing." Larry was confident I could "get up to speed" and quickly become a productive software code writer. I was thirty-one years old. Larry said I was still young and could have a lucrative career.

I did indeed become productive as a programmer very fast. For three semesters, while doing the job during the day, I attended computer science courses at Brooklyn Polytechnic Institute in the evening. These were seminars on algorithms and data structures (programming in the Pascal programming language), low-level languages (programming in x86 assembler), and Artificial Intelligence (programming in a LISP dialect for natural language processing and spoken dialogue conversational interaction with a pseudo-conscious software). Ironically, these were the sorts of classes offered at the MIT electrical engineering department in the 1970s, which I had avoided when I was there.

I read the illustrious book *Algorithms* by Robert Sedgewick.[188]

It introduced me to data structures such as arrays, linked lists, queues, and stacks.

I learned about the methods of bubble sort, quicksort, heapsort, and merge sort.

I learned about the searching methods of sequential searching, the binary tree search, the radix search, indexed sequential access, string searches, and pattern matching.

I studied the sample source code of basic yet powerful storage allocation techniques, recursiveness, run- and variable-length encoding algorithms, and parsing.

I encountered geometric algorithms, graphics algorithms, and mathematical algorithms.

I gained familiarity with virtual memory, file compression, cryptology, network flow, random numbers, polynomial arithmetic, and the Fast Fourier transform.

I worked at Neuberger & Berman in the Personal Computing Group of the Data Processing Department for three years and four months. It was a long stint. I was very successful in many ways. I got a few substantial salary raises. Above all, I learned a lot about technology.

At the end of that stretch, in July 1991, I left America definitively and for good, moving to Frankfurt with my German wife Helga.

# CHAPTER 14

Learning to be a computer programmer and becoming a specialist in certain technologies enabled me to earn a decent living for the first time in my life and in the years ahead. It was an achievement and a big change.

It was very difficult to work full-time fifty weeks a year with only two weeks' vacation, as the American system requires.

At the end of my three-plus years doing that IT job, here is the letter of recommendation that Larry wrote for me (Alan N. Shapiro private collection):

---

**Neuberger&Berman**

522 Fifth Avenue
New York, N.Y. 10036
(212) 730-7370
Facsimile: (212) 869-3419

Members New York
Stock Exchange

Investment Management

July 10, 1991

To whom it may concern:

This letter is provided to introduce Mr. Alan Shapiro, who worked in my department at Neuberger & Berman as a C Programmer/Analyst from March 14, 1988 through July 12, 1991.

Neuberger & Berman is an Investment Management Stock Brokerage firm, a member of the New York Stock Exchange and other major exchanges with principal offices in New York City.

During his employment Mr. Shapiro developed specialized systems and programs in the MS-DOS, Microsoft Windows, PRIMOS, (the operating system used by PRIME Computers), and UNIX environments. He is highly experienced and specialized in microcomputer networking, file transfer and Mini/Micro applications. He has most recently developed a customized application for Bank Loan Collateralization and has worked extensively in the development of a workstation for Portfolio Management.

In addition to his technical skills, Mr. Shapiro has always been a dedicated and conscientious employee. We here are sorry that he is leaving, but recognize his personal desire to move to Germany. I am sure that he will be a valuable employee wherever he goes. He has my very highest recommendation. Should you prefer a more personal recommendation I would be pleased to talk to you. My telephone number in New York City is 212-790-9188.

Very truly yours,

Lawrence
Director, Information Systems

---

This letter is provided to introduce Mr. Alan Shapiro, who worked in my department at Neuberger & Berman as a C Programmer/Analyst from March 14, 1988 through July 12, 1991.

## From Procedural Programming to Object Orientation

I wrote programs in the languages Assembler, Fortran, C, and C++. The book *The C Programming Language* by Brian W. Kernighan and Dennis M. Ritchie, the original definition of the language, was my bible for learning C programming.[189] Ritchie, who worked at the Bell Labs Computing Science Research Center in Murray Hill, New Jersey, created the C language.

As a C programmer, one learns about variable names, data types, constants, and operators.

One grasps the coding of program flow control mechanisms such as "switch" statements (enable you to execute different blocks of code depending on the value of a test expression) and "while" and "for" loops.

One gains an understanding of memory addresses and pointers (variables that store memory addresses).

```
main()
char* firstIntegerInput[16];
char* secondIntegerInput[16];
unsigned a, b, c, d, e;
{
    while (1) {
        printf("Enter first integer\n");
        b = atoi(gets(firstIntegerInput));
        printf("Enter second integer\n");
        c = atoi(gets(secondIntegerInput));

        if (b < c) {
            c = b; b = c; c = e;
        }
        a = b % c;
        d = 1;

        while (d > 0) {
            if (a == 0) {
                printf("Greatest common denominator: %u", c);
                d = 0;
            }
            else {
                b = c;
                c = a;
                a = b % c;
            }
        }
    }
}
```

C code that implements the algorithm of Euclid (the ancient Greek mathematician) to find the greatest common denominator for two integers.

One masters "structures" (user-defined data types that group several variables into a combined data holder) and functions (blocks of code that perform specific tasks, contributing to modularity, code reusability, and greater efficiency).

One acquires knowledge about the built-in file manipulation functions and graphics programming.

One becomes familiar with the functions available to the programmer in the standard libraries of Microsoft C, such as input and output routines (<stdio.h>) and string manipulation functions (<string.h>).

The decade-by-decade history of programming language paradigms brought successive innovations, such as the concepts of functions (1970s), event-driven programming (1980s), and the "classes" and "software objects" of object orientation (1990s). Computer programming started historically with machine, assembly, and higher-level languages. A typical program in the 1960s was the "spaghetti code" of a series of sequential instructions issued to the processor linearly inside a single "main" procedure. With the 1970s innovation of functions, program control can be delegated to helper routines. There is a reduction in the writing of duplicate code.

An input-output dialogue occurs between the "calling" and the "called" function. The "calling" function can pass data via parameters to the "called" function, to which it temporarily hands over control. When the called function is finished, it sends control back to the caller with a return value as output. Yet code and data are still strictly separated from each other. In this paradigm, technology is a tool deployed instrumentally by the human subject who is in charge, acting on a non-living machine or "thing." In event-driven programming, the program no longer proactively calculates something or sends instructions to the processor. The software sits passively in a loop. The code gets executed many times per second while waiting reactively for a user input event to occur via devices such as the mouse, keyboard, microphone, or camera. This is beyond the programmer-as-master ruling over the machine-as-"slave" model. It is a step towards the software as autonomous and semi-alive.

I learned the C++ programming language, an object-oriented extension of or successor to C. An object-oriented software object is a complex data type composed of many values of many variables grouped together and the code actions that operate on that data. Software classes are either built into the language, made available for use by third-party libraries, or written by the programmer herself. The "software class" is the specification, and the "software object" is a single runtime instance of the class. The class *encapsulates* the values of the properties (known as "fields") of an object and the operations (known as "methods") on that object into a single "object-oriented" unified entity. Data and code are unified. In this programming paradigm, the software object is on its way to becoming self-aware. An object has "introspection": it knows its internal data and its actions on itself.

While migrating from C to C++, I learned about inheritance and polymorphism. Inheritance allows a "child class" to derive behaviors from a "parent class." Class inheritance hierarchies follow the logic of generalization and specialization. Polymorphism allows a method to use variables of different data types at different moments.

Object-oriented programming assumes the existence of so-called "real-world" processes. Software development sets itself the task of modeling these processes in virtual spaces. The object-oriented software designer gathers the requirements for the system, application, virtual world, or game to be programmed. She identifies the classes that will comprise the software's framework, the responsibilities of each class, and how the classes will collaborate. The object-oriented designer thinks about the extensibility and maintainability of the code. The OO designer works with the visual tools of the "class diagram" and the "class model" to describe the system. She is guided by well-known "best practices design patterns."

```
class Book {
public:
    string myStringBookTitle;
    string myStringBookSubtitle;
    int myNumCopiesSold;
};
int main() {
    Book myBook;
    myBook.myStringBookTitle = "Venice in Las Vegas";
    myBook.myStringBookSubtitle = "An American and European Auto-Socio-Biography, 1960s to 1980s";
    myBook.myNumCopiesSold = 1000000;
    // print attribute values
    cout << myBook.myStringBookTitle<< "\n";
    cout << myBook.myStringBookSubtitle<< "\n";
    cout << myBook.myNumCopiesSold << "\n";
    return 0;
}
```

C++ code that creates an object called "myBook" instantiated from the class Book and accesses its attributes.

CHAPTER 14

## Windows Programming

*During his employment, Mr. Shapiro developed specialized systems and programs in the MS-DOS, Microsoft Windows, PRIMOS (the operating system used by PRIME Computers), and UNIX environments.*

    I became proficient as a Windows programmer. I specialized in graphical user interface coding. I developed communications systems to transfer and update real-time data between the PRIME Computers and DEC (Digital Equipment Corporation) VAX minicomputer hosts and the Windows applications running on the workstations of the well-heeled Wall Street stock-holding investment portfolio managers and bank loan decision-makers.

    I learned about utilizing the Microsoft Windows API (Application Programming Interface) via the Windows SDK (Software Development Kit). The Win API is

```
hWnd = CreateWindow("Generic",          /* window class      */
    "Generic Sample Application",       /* window name       */
    WS_OVERLAPPEDWINDOW,                /* window style      */
    CW_USEDEFAULT,                      /* x position        */
    CW_USEDEFAULT,                      /* y position        */
    CW_USEDEFAULT,                      /* width             */
    CW_USEDEFAULT,                      /* height            */
    NULL,                               /* parent handle     */
    NULL,                               /* menu or child ID  */
    hInstance,                          /* instance          */
    NULL);                              /* additional info   */


WNDCLASS WndClass;
long FAR PASCAL GenericWndProc(HWND, unsigned, WORD, LONG);
    .
    .
    .
int PASCAL WinMain(hInstance, hPrevInstance, lpCmdLine, nCmdShow)
HANDLE hInstance;
HANDLE hPrevInstance;
LPSTR lpCmdLine;
int nCmdShow;
{
    if (!hPrevInstance) {

        WndClass.lpszClassName = (LPSTR) "Generic";
        WndClass.hInstance = hInstance;
        WndClass.lpfnWndProc = GenericWndProc;
        WndClass.style = NULL;
        WndClass.hbrBackground = GetStockObject(WHITE_BRUSH);
        WndClass.hCursor = LoadCursor(NULL, IDC_ARROW);
        WndClass.hIcon = LoadIcon(NULL, IDI_APPLICATION);
        WndClass.lpszMenuName = (LPSTR) NULL;
        WndClass.cbClsExtra = NULL;
        WndClass.clWndExtra = NULL;

            if (!RegisterClass(&WndClass))
                return (NULL);
            .
            .
    }
}
```

Code from the original Microsoft Windows Software Development Kit documentation that creates a window and registers a Windows class called "Generic."

the fundamental mechanism for accessing the operating system's many features supporting the running program. I worked with the early 16-bit operating system versions of Windows 2.1, 2.11, and 3.0, also known as Win16.

## Network Programming

*He is highly experienced and specialized in microcomputer networking, file transfer, and Mini/Micro applications.*

The Data Processing Department of Neuberger & Berman used minicomputers from Prime Computer as their primary Information Technology platform. The operating system running on these servers was called PRIMOS. PRIMENET software enabled access to all files and executables across multiple machines as if they all resided on a single virtual network.

I wrote data communications programs using the standardized TCP/IP protocol suite of networking technologies for the Internet (Transmission Control Protocol / Internet Protocol). In the application layer of the framework, I wrote "Sockets" code for file transfer (FTP – File Transfer Protocol), e-mail send and receive (SMTP – Simple Mail Transfer Protocol), remote login (Telnet), and remote host management (SNMP – Simple Network Management Protocol).

```c
bool SendFileOverSocket(int socket_desc, char* file_name) {
    struct stat object;
    int file_desc;
    int file_size;

    stat(file_name, &object);

    file_desc = open(file_name, O_RDONLY);
    file_size = object.st_size;

    send(socket_desc, &file_size, sizeof(int), 0);
    sendfile(socket_desc, file_desc, NULL, file_size);

    printf("File %s of size %d bytes sent successfully over FTP socket\n",
        file_name, file_size);

    return true;
}
```

Code segment in C for FTP file transfer using TCP/IP Internet Sockets.

## Another Nasty Boss

The three-plus years I worked at Neuberger & Berman were an amazingly positive experience. I learned the practices of computer programming and software development, skills which enabled me to support myself economically for the rest of my life. Having a socially recognized profession strengthened my self-esteem. The knowledge I gained was invaluable for later realizing my dream of writing about technology and society as an "original thinker."

There was a price to be paid for all those good things I acquired. The downside of the situation was the nastiness of my immediate supervisor, Steven. After the first year, Steven became my boss in our very small four-person sub-department. He was not as dreadful as Norm at my previous job at Greenwood Mills but came close. He insulted me often and gratuitously. Steven was a nice guy in his behavior towards many other people. My German wife Helga explained that one meaning of the term "biker" in the German language is someone in the business world who "rides over" the people below him in the hierarchy while smiling, friendly, and looking up at those above. This label seemed to fit Steven well.

I am unsure what secret psychological trauma and desire compelled Steven to mistreat me. I guess it was his repressed insecurity and sadism.

> "Top 'o the morning to you, guv'nor!" he said, mockingly imitating a British accent and phrasing. "It's time for your code review!"
>
> He would reprimand me sarcastically in front of others at meetings with managers, operators, and programmers from the larger minicomputer team. "Wow, he really has you hopping!" said one of the gilded portfolio managers to me.

Steven lived a few years in Israel and was a staunch Zionist. I worked with him for two years, controlling myself never to blurt out once that I felt some sympathy for the Palestinians.

Steven's abusive conduct towards me made the job increasingly unpleasant and threatened to cause psychosomatic ailments. I could not afford to get sick. With my new informatics knowledge and five years of experience, I thought I could easily get another, even higher-paying, job. But that proved to be surprisingly tricky. I was a C language, Windows, and communications programmer. Handed a given software project's requirements, specifications, or design, I could pragmatically write a well-crafted, functioning, and bug-free program. However, when I was interviewed for a Windows programmer position at a large bank or another Wall Street firm, I fumbled when they asked me to explain conceptually how the Windows operating system is constructed. I bombed on a couple of those interviews because I did not grasp deeply how Windows is put together according to the software architect's understanding. It reminded me of my failure to "get" calculus in the twelfth grade.

After five years of living in New York and America, Helga told me she was returning to Germany – either with or without me. I could either go with her or not, it was my decision. If I decided to stay in New York, our love relationship and marriage would end.

I decided to go with Helga to Frankfurt, Germany. America, I'm outta here!

There were four reasons for my decision to leave the country at that point:

(1) I wanted to continue being with Helga and did not want to go back to being alone.
(2) It was a reawakening of my longstanding dream to move to Europe. I failed in my youthful attempts to settle in France or Italy. The idea of Germany offered a new chance. Going together with a woman made creating a home there seem more realistically possible.
(3) I could no longer stand the stress of working under Steven.
(4) My search for a new job in New York – to get away from Steven – was at a standstill.

My duties and skills at the job I had were not necessarily transferable. I was stuck. I secretly told Larry (my boss's boss who had initially hired me) that Steven was "torturing" me. I asked if he could reassign me to another working group within the Data Processing Department. Larry said no. It was not organizationally possible.

In July 1991, I gave my famous "two weeks' notice" and prepared to emigrate.

> In addition to his technical skills, Mr. Shapiro has always been a dedicated and conscientious employee. We here are sorry that he is leaving, but we recognize his desire to move to Germany. I am sure that he will be a valuable employee wherever he goes. He has my very highest recommendation.

Here is the letter of recommendation that Steven wrote for me at the end of my three-plus years doing that IT job (Alan N. Shapiro private collection):

# CHAPTER 14

**Neuberger&Berman**

522 Fifth Avenue
New York, N.Y. 10036
(212) 730-7370
Facsimile: (212) 869-3419

Members New York
Stock Exchange

Investment Management

Alan Shapiro worked as a C Programmer/Analyst under my supervision for about two-and-a-half years. Mr. Shapiro is an outstanding individual. It has been a great pleasure to work with him, both professionally and personally.

Mr. Shapiro has a truly rare combination of strengths and skills. He has a creative mind, and is able to imaginatively conceive of solutions to difficult and subtle problems. He also has a keen sense of logic, and is extremely dedicated in his attention to details.

Mr. Shapiro has a broad and deep knowledge of micro-computing. He has also demonstrated great aptitude in designing and implementing multi-environment and cross-platform systems and applications. The programs he writes are both stylistically elegant and full of breakthroughs in their application of software design principles and knowledge of machines.

While at Neuberger & Berman, Mr. Shapiro made important contributions in the areas of Portfolio Management and Bank Loan Collateralization applications, network systems development, and distributed processing design.

I strongly endorse and recommend Mr. Shapiro as a highly qualified professional in his field. He will be an asset to any organization.

Steven
Manager of Personal Computing

---

Alan Shapiro worked as a C Programmer/Analyst under my supervision for about two and a half years. Mr. Shapiro is an outstanding individual. It has been a great pleasure to work with him, both professionally and personally.

Mr. Shapiro has a rare combination of strengths and skills. He is creative and can imaginatively conceive solutions to difficult and subtle problems. He also has a keen sense of logic and is extremely detail-oriented.

Mr. Shapiro has a broad and deep knowledge of micro-computing and has demonstrated tremendous aptitude in designing and implementing multi-environment and cross-platform systems and applications. The programs he writes are both stylistically elegant and full of breakthroughs in their application of software design principles and knowledge of machines.

While at Neuberger & Berman, Mr. Shapiro made important contributions to Portfolio Management and Bank Loan Collateralization applications, network systems development, and distributed processing design.

I strongly endorse and recommend Mr. Shapiro as a highly qualified professional. He will be an asset to any organization.

A woman who worked on the trading operations floor at Neuberger & Berman heard that I was quitting to move to Germany and wanted to talk with me. She was alternately sympathetic and mocking. She shared my desire to leave America and thought Germany was cool. She had already gone to Germany herself to try to live there but could not find a job. She appointed herself the issuer to me of an authoritative "warning" that finding a job would be extremely difficult. She taunted me with the ominous prediction that I would fail to find a job and would soon be back in New York. Her "word to the wise" scared me a little, but I was laughing inside. She did not know or grasp that I was a programmer and a specialist in Personal Computers and the Windows operating system, and that those were hot skills in the early 1990s.

## The Ecstasy of Speculation

Many years later, I think back on my time in New York City, working in the financial industry and supporting data systems for Wall Street portfolio asset managers. This experience inspires some general reflections about capitalism and neo-fascism in me.

Neo-liberalism and globalization, as they were carried out, are bad things. Yet globalization is inevitable, unstoppable, and irreversible. The rise of globalization in the 1990s was, in important ways, the consequence of the intensified networking of computers, which catalyzed the virtualization of capital and the acceleration of electronic and liquid money flows.

It does not matter how many pseudo-populist, hate-inspiring, demagogic politicians – with their isolationism, nationalism, racism, and xenophobia – rail against the supposed elites of what they superficially define for their followers as globalization. Contrary to the conspiracy stories spun by these neo-fascists, globalization's essence is the free-floating circulation of money, the unleashing of high-speed virtual capital transactions that know no territorial borders, a self-reproducing perpetual motion machine seeking limitless profit wherever it can find it.

If they came to power, the self-proclaimed populists would implement a reactionary "libertarianism" worse than neoliberalism that would dismantle and/or privatize many state functions (environmental regulations, consumer protection, worker safety, public education, health management measures, social security retirement funding, etc.), thus giving even freer rein to the wild propagation of speculative capital. We would migrate from so-called globalization to an even more radical and destructive version of the post-industrial economy.

Speculative capitalism is the successor to industrial capitalism. The "ecstasy of speculation" (Achim Szepanski) tends towards fictionalization, detached or abstracted from the production and distribution of tangible physical goods and services.[190] Speculative capital operates in the interconnectivity spaces of the vast webs of information and so-called communication. It thrives in the atmosphere of the manic "attention economy" of celebrities big and small, platform and surveillance capitalism's acquisition of troves of "personal data," the "post-truth" system of discursive power and Artificial Intelligence algorithms, and the image-intensive simulations and simulacra of the media culture. This system is open 24 hours a day, every day and night of the year.

There is a promiscuity of all exchanges. Capital moves about in a parallel universe. Finance and the cyber-techno-grid have become Earth's satellites and have gone into orbit. Internet memes and tropes spread through the matrix because they are contagious or infectious, as evidenced in phrases like "going viral" and "viral media."

Postmodern culture is saturates with pornographic imagery, showcasing an endless stream of "shocking" or voyeuristic images of every possible "erotic" minutia, which no longer shock. Media culture supplies an endless flow of graphics and words, a universal availability where nothing is hidden. There is a homology between the hyperreality of images and the catastrophe of verbal deception in political discourse – and they reciprocally affect each other.

With his concept of "the spectacle," the media theorist Guy Debord understood that the omnipresence of visual images institutes a world of passivity, a diminishing of what is "directly lived," and an increase in the autonomy and power of the images themselves.

Debord's book *The Society of the Spectacle* develops an analysis of the post-Second World War advanced capitalist society: the ubiquity of the mass media, high-tech, the culture of images, television, movies, advertising, computers, consumerism, marketing, organized leisure, shopping malls, cybernetics, circulating transport and telecom networks, and the tourist industry. The book's main thesis is that we are fundamentally positioned as spectators in the media and consumer culture. In his later book *Comments on the Society of the Spectacle*, Debord begins the development of the powerful concept of the "integrated spectacle," describing a much more sophisticated and pernicious stage of the subtly totalitarian society of control.[191]

My years as a Programmer/Analyst at the Wall Street stock brokerage firm Neuberger & Berman were my first brush with the global financial system. It would not be my last.

CHAPTER 15

# Venice in Las Vegas

America is a system of cultural models and codes, omnipresent images and advertising, ubiquitous automobile circulation, telematic networks, self-referential rhetoric, spectacles where the performing individual is fully integrated, and hyperreal simulations that "precede the real." *The real* is the "sociological reality" in which sociologists believe. Mainstream sociologists base their "scientific sociology" on a nineteenth-century paradigm (that of Auguste Comte), which assumes a world of docile objects waiting to be "objectively" investigated. It is a classical pre-quantum physics "scientific" worldview that assumes a "social world" rationally ordered by the sovereign thinking Cartesian subject of social science who is in control.

An alternative literary-existentialist sociology might also be scientific. It would be based on the hyper-modernist natural sciences of quantum physics, relativity, and chaos/complexity theory. Literary sociology would consider stranger and more wily objects in a never-to-be-mastered social field of radical uncertainty and paradox. Everything is enigmatic and aleatory.

## The Secret Affinity Between Gambling and the Desert

> "The intensity of gambling reinforced by the presence of the desert all around Las Vegas. The air-conditioned freshness of the gaming rooms against the radiant heat outside. The challenge of all the artificial lights to the violence of the sun's rays. Night of gambling sunlit on all sides; the glittering darkness of these rooms in the middle of the desert. Gambling is a desert form, inhuman, uncultured, initiatory, a challenge to the natural economy of value, a crazed activity on the fringes of exchange." (Jean Baudrillard, *America*[192])

The desert and gambling are finite, concentric spaces that grow in intensity towards their interior, towards a central point. It is a space of predilection, a space where money loses its value in the casino, where there is an extreme scarcity of traces and a lost shadow in the desert.

Gambling in the context of the American simulacrum is neither purely an entertainment activity nor purely an activity pursued with the hope of making money without effort. It is both. Gambling is a paradox. It is an entertainment activity that promises to make money without work as one of its key elements of attraction. Its appeal to the player is the tension between these two aspects – entertainment and easy money.

The player never believes that she frequents the casino to make money. If she did, she would not ritualistically persuade herself that her expenditure is the price of a legitimate consumer activity in which she has the right as a "media culture citizen" to indulge periodically. Gambling's ambivalent structure is the "quantum physics *complementarity paradox*" between the packaging of an entertainment experience and the illusion of instant monetary gain.

Consider the "doublethink" (George Orwell's *1984*) or "double bind" (anthropologist and cyberneticist Gregory Bateson) of the cognitive processes of the player.

Orwell defines "doublethink" as "the power of simultaneously holding two contradictory beliefs in one's mind and accepting both."

For Bateson, the "double bind" in psychology occurs when a person simultaneously receives two or more contradictory communications or signals.[193]

The gambler knows the "objective reality" of the casino: the house advantage, the management's profit calculation, and the laws of probability. However, she is not deterred from playing because she receives symbolic messages from the architectural and cultural design of the casino that reinforce belief in "personal exemption" and "beating the odds." The player rationalizes that the sequence of wagers and outcomes will pass through a different "causality" than probability. Each player believes that she will evade the inexorable logic of house profit.

This "doublethink" is analogous to people's attitudes toward social structure, economics, and powerful cultural models and codes generally in their lives. Although I know that a society with definite organizational patterns exists, I detach myself from this awareness and emphasize, in my mind, my autonomy as an alleged individual, my decontextualized self. Instead of thinking about capitalist social structures and codes of oppression, exploitation, and alienation, I dream of "getting out of them" by winning in the casino or winning the lottery!

## The Total Design Environment

Casino gambling is an embodied metaphor for our situation in hypermodern society and the American Way of Life in the age of Donald Trump. The "simulacrum strategy" of casino management is to immerse the player in a highly controlled semantic and semiotic environment. The *semantics* is the management's design of a controlled ambiance and selling a packaged consumer experience to the players. The *semiotics* is the transformation of the value of money inside the casino. Money becomes chips. The player is deprived (to some extent) of the awareness that she is playing with real money.

Inside the casino, the difference between day and night is eliminated. The same activity continues uninterrupted twenty-four hours a day. No clocks are visible anywhere in the casino. There are no windows, creating an architectural impression of limitlessness. The casino often consists of one enormous room the size of a football or soccer field. You cannot see the other end of the room upon entering. Abundant mirrors create the effect of an infinite refraction. The number of columns is minimized, giving the impression of an entirely suspended ceiling.

The design environment is even more encompassing for the gambler who stays at the adjoining hotel. Everywhere, there are shops, comfort facilities, and personal services. The gambler does not need to seek the "satisfaction" of her "needs" anywhere else. The valuation of the consumer experience of losing money at blackjack, roulette, slot machines, video poker, or sports wagering is enhanced by the total design ambiance where other commodities, spectacles, and semiotic signs of "the good life" are on prominent display.

## The Simulacrum is More (Hyper-)Real Than the Original

The copy or simulacrum has replaced the original in the architecture and ambiance of Las Vegas. Many of the hotel-casinos are simulacra of specific histories or geographies.

The simulacra of "the good life" substitute for the good life.

At "Caesar's Palace," the semiotics of ancient Rome replace the real historical Rome.

The Eiffel Tower at the Paris Las Vegas Casino replaces the Eiffel Tower in Paris.

The skyline of skyscrapers and the Brooklyn Bridge at the *New York-New York Hotel and Casino replaces Manhattan.*

It was in Las Vegas where I was able to relive all of my essential European and New York wanderings.

New York City in Las Vegas. Duplicates of Christopher Street and Sheridan Square in the West Village. The entrance to the numbers 1 and 9 subway lines. Replica of the Statue of Liberty. Copies of the Empire State Building and the Chrysler Building.

CHAPTER 15

New York-New York Hotel and Casino, Las Vegas (Photos by Alan N. Shapiro)

Donald Trump's Taj Mahal (in Atlantic City, New Jersey, the East Coast twin of Las Vegas) replaces the ivory-white marble mausoleum in the Indian City of Agra.

The replicas of the *Campanile* (Bell Tower) of St. Mark's Church and the Rialto Bridge at the Las Vegas Venetian Hotel Casino replace the real Tower and Bridge of the city of Venice.

The simulation of Venice in Las Vegas, its cultural-imaginary presence in the former desert of the former territory of Nevada, down to the details of gondolas, canals, and bridges.

Bell Tower of St. Mark's Church, Venice (left) and Las Vegas (right) (Photos by Alan N. Shapiro)

Shylock in *The Merchant of Venice*. The famous Venetian bridge called the Rialto was a hub of antisemitism in Shakespeare's time. Today's Las Vegas Venetian Hotel-Casino has a nice Jewish deli called *The Rialto*.

Shylock: "Yet his means are in supposition: he hath an argosy bound to Tripolis, another to the Indies; I understand moreover, upon the Rialto, he hath a third at Mexico, a fourth for England, and other ventures he hath, squandered abroad."[194]

The simulation of Venice's Rialto Bridge: cars rather than water run underneath. In the inner city of the "real" Venice in Italy, there are no cars.

Rialto Bridge (left) and Rialto Deli (right), Las Vegas (Photos by Alan N. Shapiro)

The architectural reproduction of other parts of the world in Las Vegas is an elaborate three-dimensional mirage devised to distract the gambler from the fact that he is being separated from his money. The gambler loses his money in order to win the (copy of the) world.

After a good night's sleep at the Marriott Residence Inn on Paradise Road near the Las Vegas Convention Center, we drive to Las Vegas Boulevard South and park the car on level 8 of the parking complex. We descend the elevator *that is about to transport us to Venice, Italy*.

The simulation architecture and ambiance exceed the intentions of the casino hotel owners. *Learning from Las Vegas* (a famous book about "postmodern" architecture by Robert Venturi, Denise Scott Brown, and Steven Izenour) today is to learn that Venice is, more than anything else, an idea.[195] Paris is an idea. New York is an idea. An idea can be expanded and extended. Venice is not restricted to its physical location in the Veneto region of northern Italy. Embodied semiotics is what makes "reality." I am physically in Las Vegas, but I am "really" in Venice. "Culture has never been anything but the collective sharing of simulacra." (Jean Baudrillard, *Fatal Strategies*)[196]

It's time for lunch. The architecture and décor of the Grand Lux Café on the ground floor were inspired by the look and feel of classic Venetian cafés. I will have the Quattro Formaggi or "Four Cheeses" pizza *per favore*: mozzarella, parmesan, roman, and fontina.

I love it when they replace something that was authentic and historical with a pictorial simulation of itself. They did that with the famous upper deck outfield façade at the old Yankee Stadium (baseball) in the Bronx. They did that with buildings in Frankfurt's main square (the Römer) that were destroyed during the Second World War. "Old-fashioned-historical-looking buildings" stand now in their place.

The term **simulacrum** – an image or semblance of something – derives from the Latin *simulare*. Starting with Plato, Western philosophy has regarded what it calls the simulacrum with suspicion. Thinkers associate the simulacrum with falsity, implying a dualistic opposition between truth and simulacrum. For Jean Baudrillard, the simulacrum is what is "true." The simulacrum conceals the state of non-existence of conventional "truth."

The idea of reality was already a cultural construction of Western civilization. Our notion of "the real" was always already a simulacrum. This is what makes "the virtual" possible. "Reality" in our culture was always an illusion. This chimera was maintained by the demarcated difference between "the real" and representation. The media culture breaks down that difference. The proximity of so-called "reality" to the models and codes that instantiate it and to the image-copies that purport to leave the privileged status of the originals intact leads to corruption. Reality and the image move into each other's spaces.

The definition of "the real" in hypermodern culture is that of which it is possible to make an equivalent reproduction.

Jean Baudrillard's most celebrated book is his 1981 volume *Simulacra and Simulation*, where he famously wrote about the map preceding the territory and about Disneyland existing to conceal the fact that all of America is Disneyland. He writes:

> Disneyland exists to hide that it is the 'real' country, all 'real' America that is Disneyland (a bit like prisons exist to hide that it is the social in its entirety, in its banal omnipresence, that is carceral). Disneyland is presented as imaginary to make us believe that the rest is real, whereas all of Los Angeles and America surrounding it are no longer real but belong to the hyperreal order and the order of simulation [...] The imaginary of Disneyland is neither true nor false, it is a deterrence machine set up to rejuvenate the fiction of the real.[197]

Disney exists to save the "reality principle" or the myth of an "authentic real."

American culture is hyperreal in its belief that making a copy is the ultimate certification of originality. The simulation process transforms so-called "reality" into an inferior version of the imitation. Hyperreality henceforth rules. In an inversion, the copy becomes the model to which the original must answer. The latter pales in its "graphic resolution" compared to the former. Copying paradoxically captures the authenticity of the original.

Simulation is the process of substituting the signs of the real for the real. Simulacra are the artifacts, like the replicants in *Blade Runner*. Semiotics (linguistics applied to culture) teaches about "the signifier" and "the signified," which together constitute the linguistic-cultural sign. In postmodernism, the signifiers (images and discourses) come to replace the signifieds (facts and references) of which the visuals and words were supposed to be the reliable and verifiable representations. Representation, which privileged imagination as the healthy distance between fiction and the real, is superseded by simulation, where the salutary gap between reality and virtuality has disappeared. Words and images come to stand on their own and have no basis in "facts." They are copies without originals. The mythical original becomes a mere "reality effect."

Traditional social solidarity disappears and then reappears as a simulation. Media culture produces endless simulations of the absent social space.

Instant winning. Instant wealth. Match two halves of the coupon and win five hundred dollars! Anyone can become an instant millionaire or billionaire by winning the lottery or hitting the jackpot! Anyone can play Wall Street.

Money has become increasingly necessary for survival and satisfaction. People have become money-crazed. Speculations like the stock market and gambling assume center stage in hypermodern society.

## The Sexuality of Gambling

**Blackjack.** Each deal of the cards, each game incident, is independent of all others. The outcomes of incidents that have already occurred have no influence on the incident at hand. The chances of winning before the cards are dealt are the same each time. I assume here the ideal condition of a fresh deck. The cards already dealt in a partially played deck cause some alteration of the odds.

Streaks, both lucky and unlucky, only exist retrospectively. There is no way for the player to know when in the middle of a streak when it will end.

One of the fascinating aspects of the game is its paradoxical dualisms. One of these "quantum complementarities" is that each round of play is both a ritual and a unique event. The "bad faith" (Jean-Paul Sartre) of the player is his imaginary unification into a narrative (in his mind) of the isolated instances of the hands of cards or spins of the wheel.[198]

There is a hyperreal sexuality on display in the casinos that complements their airport-like nowhere-space design and futuristic interior motif. It is the functionalized sexuality of cocktail waitresses and female croupiers in mini-dresses. It is the simulation-performance of sexuality promoted by the media-commodified erotic system, with its heteronormative male/female binary and its stylization of the ultra-feminine. This system is deeply misogynistic and objectifying of women. This operational eroticism is part of the simulacrum of opulence and "polymorphous perverse" paradise with which the management surrounds the player.

The eroticism of gambling can be compared to the striptease artist's evocation of desire (see Roland Barthes, *Mythologies*[199]). The audience for the striptease is captivated by auto-erotic gestures that radically exclude the observer: self-caressing and a style of dancing for apparent self-satisfaction. The girl (or other erotic artist) is inaccessible. Her gaze is intended for no one. This is her attraction to the voyeuristic viewer. In gambling, there is a similar exclusion of the self. The number of decisions I make is limited. The white ball spins around the wheel. The cards at blackjack nearly play themselves. I am intoxicated by the variations of the cards, the unfolding of their finite number combinations and sequences. One reduces the house advantage to a minimum by sticking to the monotonous perfection of a mathematical formula. One makes the same gesture of drawing a card or standing pat in each possible situation according to a system that repeats itself forever.

We experience the systematic perfection of a pointless activity. We temporarily escape the tensions of daily life and enter a realm of play with rules and ideal conditions marked off from "the real." The risk involved when there is a wager intensifies the element of fantasy.

The autoeroticism of a space-age slot machine is the solipsism of the device iterating through all its permutations, observed by the human repeatedly pulling the electromechanical arm or pressing the same spin button over and over, like the rat in the lab experiment who presses the same lever, again and again, to get more stimulation sent directly to the pleasure center in its brain – until it finally expires.

## Online Gambling

Today, in the era of digitalization, with widespread online gambling, the distinction between casinos –where gambling takes place – and the rest of the world, which is "not a casino" – where gambling supposedly would not take place – has disappeared.

The online casino employs webcams and remote telepresence to enable me to be "virtually" at a physical casino's roulette and blackjack tables with croupiers and dealers but no physically present players. The online casino where I gamble is in Latvia and Georgia. The "game presenters" are mostly young Russian and Latvian women and a few young British men. I can engage in conversation with a croupier while playing roulette or blackjack. I hear what they say back to me and other players through my speakers or headphones. I make my comments by typing into a chat box.

Online casinos can teach much about the virtualization of money in the contemporary age of financial speculation, the circulation of floating capital, the endless cycles of bubbles and crashes, and the decoupling of money from "gold standards" and "production of wealth."

Watching a college football game on TV shows how the game itself has been subordinated to the cult of information, the trans-media convergence with the Internet, and the instantaneity mania of the gambler. Like on the World Wide Web, the football game is now just a video "window," taking up only a portion of the television screen. At the bottom of the screen is a nearly permanent band of information reporting on the up-to-the-second scores of all other ongoing games. The rest of the screen is occupied by statistical windows with transparent backgrounds, displaying information pertinent to the game currently streamed in the video window. The intrinsic love of the game has been "downsized." The gambling viewer is less interested in the drama of this particular game than in the instantaneous informational updates on the games on which he has placed his multiple and parley bets. He is enthralled by the split-second ups and downs of his wagered prospects. Technical "special effects" also virtualize the game's moving images in relation to the physical screen: scenes of players are flipped sideways, spun diagonally into the background, or whisked off to oblivion.

Horse racetracks have introduced satellite networks to their equine environment. The Intertrack Wagering System (ITW) is a "simulcast" system of TV transmissions of live racing from "server" racetracks to multiple "client" racetracks that broadcast the physically remote races, offering them as additional wagering events. Wagering on televised races has so greatly exceeded wagering on the "real" races that the tracks end up having increasing difficulty attracting horse owners to participate in their local non-mediatized races. Horse racetracks have also added massive video slots and video poker salons to their premises.

Is the event henceforth a mere excuse for the enactment of the gambling passion? The synthetic event of the permutations of a computer program operating the internals of a video gambling game increasingly replaces "real" events. An Artificial Intelligence "game presenter" replaces the human blackjack dealer. I am at my gambling terminal, perpetually re-calculating the mutable outcomes. I, the human, am a component of the computer program. I exist in the interstices of the network. Soon, my real-time emotional states will be observed and analyzed by facial recognition and affect classification, data-feeding the logic and next-step moves of the Deep Learning artificial neural network running the game.

If I make a large number of small bets, the system will function perfectly, and I am guaranteed to lose. I have a better chance if I bet an enormous sum on a single game or outcome.

One obvious explanation for the "gambling explosion" in American society is the decline of the belief that one can "get ahead" through hard work in the contemporary economy. Ultimately, a random statistical system could decide who becomes rich.

Baudrillard writes in *Fatal Strategies*: "It is not enough for theory to describe and analyze. It must itself be an event in the universe it describes. In order to do this, theory must partake of and accelerate this logic."[200]

## Whiskey Pete's and the "Chance" Event

Moments after entering Nevada by car, we encountered Whiskey Pete's casino. Located at the California-Nevada state line in Primm, Whiskey Pete's hotel and

Whiskey Pete's Casino, Primm, Nevada (Photo by Alan N. Shapiro)

casino is thirty-five miles from downtown Las Vegas. Here, a wild Baudrillard conference took place in 1996.

It was dusk as I walked across the parking lot, my right jeans pocket loaded with American quarters (twenty-five cent coins) that I needed to sacrifice to the slot machine gods. I want to win some money right now!

I knew I was standing at this spot for probably the only time in my life. The first and the last time. "No two moments of your life are exactly alike" — the clichéd sentence from some book on consciousness and spiritual growth that I had read rose in my mind. But I am a materialist. I want to win some money right now!

In her essay "Chance: A Philosophical Rave in the Desert," the novelist and performance artist Chris Kraus writes about the Baudrillard conference at Whiskey Pete's Casino that she organized in 1996.[201] Baudrillard himself was the star performer at "The Chance Event."

Chris Kraus writes:

> [It would] be no mere academic conference, no recitation of concepts [...] It would be a highly structured, chaotic field with equal time for philosophers and gamblers. DJ Spooky gave the 'Keynote Address' as an ambient trip-hop performance. We had Diane di Prima, the legendary Beat poet, and her partner Sheppard Powell, I Ching divinator. Doug Hepworth traded stocks on Wall Street using chaos theory. Mike Kelley created Chance's neo-psychedelic poster and formed The Chance Band, which played backup to Baudrillard's vocals. Since the casino had been built on Indian land, I sought out Calvin Meyers, a nuclear-waste advocate for the Moapa Band of Paiutes [Native Americans of the Southern Tribes]. Meyers led participants on a Sunday morning desert walk.
>
> Noise bands Ohm-A-Revelator and Towel came from San Francisco, poet band Homer Erotic came from New York City, and the Butoh [Japanese dance theater] company Renzoku performed. We had a professional croupier named Nick Kallos from the Las Vegas Gambling Academy. Rollerblader/mathematician Marcella Greening spoke on chaos theory. Mexico City poet Luis Bauz performed an homage to Baudrillard. Liz Larner performed *Learn To Deal* about her experience as a casino dealer. For the grand finale, [a member of] the Chance Band found a box of gambling chips backstage and hurled them at the audience, who erupted in an ecstasy of Free Money.

## The Goddess Fortuna

Of all the gods of Ancient Rome, only the goddess Fortuna survived as a significant presence after the advent of Christianity and into the Middle Ages. References to this pagan deity abound in the literature of the Medieval period and the Renaissance.[202] She is alluded to and portrayed by literally hundreds of authors. She was the subject of cults and held to influence the destiny of soldiers in combat, sailors at sea, amorous couples, statesmen, and wanderers.

Fortuna was inconstant and fickle. She represented the disorderly and unruly side of life. She offered men a cornucopia but took pleasure in alternately exalting and debasing them in her domination of their fate. She was envious and could deprive one of prosperity even after it had just been attained. She was depicted as having two faces: one beautiful, the other ugly, or with one eye weeping and the other laughing. It was said that her right hand was good and her left hand evil, or that she was blind (or blindfolded) and had no regard for merit. She was compared to the changeability of the moon and the slyness of a snake or serpent. He who placed his trust in her was a fool. He might tumble from even the most secure perch of honor or happiness. The inconstancy of Fortuna was the opposite of the fixity of astrology, which maintains that the stars are arranged according to a plan of primal and eternal destiny.

The regard for Fortuna as the personification of disorder was shared by followers of both astrology and orthodox Christianity. Philosophers like St. Augustine and St. Thomas Aquinas endeavored to prove that Fortuna was evil or simply did not exist. In later Italian literature, Fortuna co-exists with God but is opposed by the forces of reason and prudence. For the Renaissance writer Francesco Guicciardini, there is a struggle between Fortuna and Virtù – the latter being the only resource capable of combating the dreaded goddess.[203] Machiavelli's *Prince* wrests his triumph from the forces of the obstacles presented by Fortuna.[204]

The modern idea of "chance" – at least as it is lived inside the casino – is a reversal of the Medieval or Renaissance representation of Fortuna. In the temples of consumer risk, there is an emphasis on abundance and the luxuries which "Lady Luck" might bestow on us. The ostentatious display of the ambiance beckons us invitingly to an easily accessible simulation of Fortuna. The authors and mythologies of the past stressed Fortuna's reluctance to share gifts and the remoteness of her treasures. Her home was an uncharted island or atop an unscalable mountain. Only the most adventurous would dare to approach it. It was said in mythology that "fortune aids the bold." To conquer the lady, one must be courageous and curse or defy her.

## Riding the Lucky Streak

The player who is ahead does not quit. He believes himself to be capable of winning much more. He should take his winnings and run out of the casino as fast as possible, but the seers of fortune keep him glued to his chair. It is *he* who attracted the fortuitous cards or is in the hands of Lady Luck. He swims in the magical dream of his rendezvous with destiny.

Blackjack is an activity without accumulation. Nothing accomplished through any single draw of the cards aids the player in subsequent draws. The Frankfurt School critical theorist Walter Benjamin compares this "fragmentation" of gambling instances to the fate of the laborer in modern capitalist society, whose previously creative activity has been divided into small, isolated units.[205] Each worker is reduced to repetitively performing the same operation and is cut off from any emergence of meaning or satisfaction. With its monopoly of conceptual knowledge, management reconstructs the total work process as a simulation.

For the blackjack player, perceiving the series of gambling incidents as a unity outside or beyond the rational laws of chance is thrilling. The player believes in streaks. He believes he is "hot." The cards are falling his way. He increases his wager. In his mind, he narrates an imaginary unification of an unconnected and gratuitous sequence of events. He bestows anthropomorphic qualities on the inanimate cards, to whom knowledge of what has occurred before is attributed. The gambler who believes in streaks does not fare as well as the factory overseers with their Frederick Taylor-inspired "scientific management." He wishes the cards to be his proletarians. He ascribes to them a quality already defunct in capitalist work organization.

Although a poor decision by another player does not affect the probability of my outcomes, most players get up and leave a table where there are one or more inexperienced players. There is a widespread belief that poor play by others can "ruin" one's own chances.

## From Fyodor Dostoevsky to Paul Auster

The gambling madness described by the great Russian writer Fyodor Dostoevsky was a radical act of desperation and social exclusion. It was the risking of an entire inheritance. The protagonist of the novella *The Gambler* is ruined by his passion for the tables.[207] He is cast out by society and denied the love of Polina Alexandrovna, the woman who is the object of his longing. Reaching his friend's security and warmth in Switzerland would mean leaving the roulette wheels of the fictional Roulettenbad (based on casinos and spas in Wiesbaden and Bad Homburg near Frankfurt). The salient point about Grandmamma's subsequent gambling away at roulette of her enormous estate is that her wealth was hereditary and aristocratic, not acquired through commercial endeavor or self-activity. Today, gambling is no longer an act of social exclusion but rather of consumerist leisure. Advertising for the new gambling industry seeks to legitimate wagering as an activity no more harmful or immoral than seeing a movie.

## CHAPTER 15

For the house, everything is calculable in advance. The laws of chance are always on its side. Every night, the house will realize a guaranteed profit (after subtracting the costs of casino construction, taxes, employees' salaries, utilities, etc.) based on its "edge" in the games: four percent advantage on craps, seven percent on roulette, nine percent on blackjack, and twenty-one percent on slot machines. But, as the great novelist Paul Auster demonstrated, there is also the "music of chance."[206] The games have rules that must be obeyed. These rules are even stricter than the law, where a guilty verdict can sometimes be forgiven.

In Paul Auster's novel and film, *The Music of Chance*, a down-and-out poker player and his existentially in-crisis financial backer overestimate their skill at the game and end up in long-term enslavement in a futile attempt to settle the debt from losing that can never be repaid.

Varieties of speculative exchange dominate the backstory of the narrative. The main character, Jim Nashe, had inherited a substantial fortune of nearly two hundred thousand dollars from a deceased father he had not seen since he was a toddler. The absent parent made his money taking risks on the stock exchange. During a despairing and absurdist one-year road trip around America, the protagonist fritters away his newly acquired liquid assets on hotels, food, and gas for his car. Flower and Stone, who are Nashe's and inveterate gambler Jack "Jackpot" Pozzi's opponents in the decisive poker game, won twenty-seven million dollars playing the lottery. They are now retired and pursuing self-indulgent hobbies while living in a fancy yet rundown mansion in the fictional locality of Ockham, Pennsylvania.

Nashe and Pozzi lose the fourteen thousand dollars Nashe had left, Nashe's car, and an additional ten thousand, which they do not have and cannot pay. They are imprisoned on the grounds of the estate of the multi-millionaires, surrounded by barbed-wire fences, and forced to carry out the physical labor of building a wall in Middle Ages style with ten thousand stones imported from the ruins of an Irish castle. Despite some begging, Flower and Stone refuse to let the two "heroes" off the hook for their debt. The rules of the game, gaming, and gambling must be observed correctly until the end. Strict obedience to the rules ultimately leads, in Auster's poker novel, to a duel for life and death.

After months of grueling work while living in a trailer in the meadow of the sprawling property, Nashe and Pozzi believe they have fully paid off their debt. Their overseers then inform them that they still owe more than three thousand dollars for food, beer, tobacco, a broken window, a one-time party, and a special visit by a prostitute. They must continue their seemingly endless Sisyphean labor. Pozzi tries to escape but is caught and beaten to near death by the supervisors of his indentured servitude. He is taken to a hospital, where he probably succumbs to his wounds. Knowledge of Jack Pozzi's fate is hidden from his friend Jim Nashe.

Paul Auster writes in *The Red Notebook*:

> The following year, I was offered a job as a caretaker of a farmhouse in the south of France [...] My friend's legal troubles were well behind her, and since our on-again-off-again romance seemed to be on again, we decided to join forces and take the job together. We had both run out of money then and without this offer, we would have been compelled to return to America, which neither one of us was prepared to do just yet [...]
>
> The worst moments came for us in late winter and early spring. Checks failed to arrive [...] little by little, we ate our way through the kitchen's stockpile of food. In the end, we had nothing left but a bag of onions, a bottle of cooking oil, and a packaged pie crust [...] Given the paucity of elements we had to work with, an onion pie was the only dish that made sense [...]
>
> We went outside for a brief stroll, thinking the time would pass more quickly if we removed ourselves from the good smells in the kitchen [...] We drifted into a deep conversation about something[...] By the time we entered the house again, the kitchen was filled with smoke. We rushed to the oven and pulled out the pie, but it was too late. Our meal was dead. It had been incinerated, burned to a charred and blackened mass, and not one morsel could be salvaged.[208]

Baudrillard writes in his book *Seduction* (1979), "To enter into a game is to enter a system of ritual obligations."[209] By choosing to participate in "the rule," one is delivered from "the law." The latter turns out, paradoxically, to be less binding. Employing Baudrillard's vocabulary, one can say that, in gambling, money is *seduced*. Money is deflected from what was assumed to be its truth. Gaming diverts money from its everyday function and is the exemplary activity symbolizing all the late capitalist systems that replace the disappeared "social" and supersede the classical modernist mode of capital that was tied to real material wealth and production. The stake or wager in gambling is no longer an investment. It is not capital as we have known it. It is the newer form of capital: a pure and arbitrary exchange. What, indeed, is *the stake*? The stake is – literally and metaphorically – a *challenge* to the dominant economic and cultural system.[210]

Yet the gambling industry only partially offers the player a seat at the table of genuine challenge, an invitation to a situation of authentic risk. The casino is also a *simulation* of challenge or risk. Similar to what Baudrillard wrote about Disneyland – Disney exists to persuade Americans that the rest of America is "real" and not of the order of the hyperreal and ubiquitous fantasyland consumerist simulations – casinos exist to convince Americans that the rest of the social and economic system has not undergone a historical paradigm shift process of gamification. We would be led to believe that capital and money are still "real."

Casino gambling exemplifies what has happened to "value" in hypermodern society because it is at the crossroads of economics and culture. Atlantic City's decaying boardwalk and burnt-down piers bow before the onslaught of advancing media-consumer culture. The simulation of wealth displaces the imperfect beauty of impoverished yet real history.

In sweepstakes, contests, and lotteries, the participant has only an infinitesimal chance of winning but is enticed into spending a few dollars to have some chance rather than none. The situation parallels one of the typical marketing strategies of large national corporations: convince every "household" in America to disburse a few dollars per week on a particular product (toothpaste, deodorant, remedy to soothe an upset stomach), and you've got an enormous profit. The pattern of money flow, as small sums from many buyers to the big corporation, ensures that few notice they are paying a price higher than the product is worth.

The lottery is the inversion of fear or domestication of negativity. Random statistical events in modern society tend to be regarded with apprehension: automobile accidents, heart attacks, violent crime, or becoming the victim of a serious disease. By turning a random negative statistical event into something potentially positive, the lottery encourages our acceptance and "making the best" of the status quo, habituating us not to question the structural underpinnings that cause crashes, accidents, and calamities.

## Last Stop Las Vegas

In the summer of 1991, at age thirty-five, I was finally ready to leave America for good. I had reached the end of my long preparations, my roundabout backward and forwards, my complex, intricate paradoxical topology. Of course, I would keep my U.S. citizenship and my American cultural identity. I would figure out how to watch MLB baseball and NFL football games. I would vote for the Democrats for President and in federal elections. I would file my IRS Income Tax Return every year. But I would go to live for a very long time in Europe, the land of the free and the home of the brave.

From my years of working as a programmer on Wall Street, I saved twenty-seven thousand dollars, much more money than I ever had before. I had software developer skills that would enable me to get work overseas. I had my German wife. I had friends in Frankfurt. There was an apartment there waiting for us. Some money would go to setting up the flat, but there would be more than enough left to start our new life. We were all set.

Before flying from New York to Frankfurt, I flew to Las Vegas. It was a sort of pilgrimage – to say hello – or goodbye – to my own personal Mecca. I played blackjack and roulette for twenty-four hours straight with no sleep. I paused a few times for meals and drank a lot of coffee. I started my wagering with small amounts, played conservatively, was very lucky, and ultimately won four hundred dollars. When I got ahead a few hundred, I increased my bet. I got ahead a thousand, started losing it back, and then had enough good sense to quit and leave the casino before losing it all. That's how a gambling session usually goes. Either you lose all the chips you bought in for right away, or you have a lucky streak at the beginning, get substantially ahead, foolishly believe you are invincible, don't have the presence of mind to stop, and end up defeated. On this occasion, I broke off before I crashed out.

## Close Encounter with the Wormhole

Later, it was the year 2000. Nine years had passed since I left America for good and feathered my nest in Germany. I traveled once a year to my home country, each time for two weeks. At a certain moment, I found myself in the newly opened "Simulacrum of Venice in Las Vegas" casino: the Venetian Resort Hotel Casino, located at 3355 South Las Vegas Boulevard.

I stood for a long time at the water's edge of the outdoor replica of a Venice canal they had architected and built, complete with gondolas. Gazing into the transparent pool, I saw now, inevitably, the other end of the wormhole. I had sighted the strange astrophysical phenomenon years earlier on two occasions from its opposite open "mouth" in the "real" Venice.

I knew it was a moment of significance and even a day of reckoning, the universe settling its accounts with me. It was a rendezvous with my fate. There was no way to avoid it. By now, thanks to my Gestalt therapy, I had achieved a firm psychological grounding for my creativity and rebellion. I knew that the wormhole was not "real." I did not jump into the water in search of it. My clothes stayed dry.

The "theoretical physics" link between two remote points in space – Venice and Las Vegas, symbolic, in my auto-socio-biography, of Europe and America – existed only as a figment of my imagination. The "science fictional" connection between two different moments in time (me in Venice at age twenty-one in 1977, or age twenty-four in 1980, and Vegas in the early 1960s) was a reverie, a pie or eye in the sky, a high castle in the air, a fool's paradise.

Despite being a mind trip, what I saw while I peered through the wormhole had much meaning, as it was now on the self-reflexive or self-aware border between real and fiction. Laughing at the hallucinatory status of my vision made it more resonant. The early 1960s (glimpsed as Las Vegas) were the time of my childhood and the "scene of the crime" (as I called it in Chapter Two of this memoir) of my skipping two years in school and its consequences. Why had I seen that specific time bend, that juncture of history and biography?

In the dreamtime, in the desert, I stood before the tribunal. "Alan, we congratulate you," said one of the Magistrates. "You have arrived at a privileged secret courtroom. Now you can choose. You can return to the year 1963. You are seven years old. Do you wish to remain with children your age? Or do you wish to skip twice?"

I was silent for thirty seconds. Then I spoke: "I wish to skip twice."

Then it was done. My life as it is, my journey, is the best of all possible worlds.

At age thirty-five, I closed my apartment in Sunnyside, Queens, New York. I went with Helga to Germany. I succeeded in escaping America. We shipped our possessions to Germany in eighteen large containers. Helga still had some affairs left to settle in New York, and she was going to arrive two weeks later. I flew alone on Pakistan Airlines. Our new apartment in the Bornheim neighborhood of Frankfurt still needed to be painted before we could move in. Helga had close friends in that same part of town who had a spare room in their flat where I could stay for a while. After arriving by taxi from the airport, I unpacked the clothes, books, and other items I had with me. I had a small porcelain dog Helga gave me as a gift. Before the first night, I sat my dog on the night table next to the bed. I planted my stake in Europe.

# EPILOGUE

From three decades into the future, I am happy to report that "all's well that ends well." I survived and lived and prospered. I turned out well and came out on top. All the unrealized fates I sampled before my thirty-fifth birthday – academic, gambler, philosopher, computer programmer, traveler, baseball fan, lover, vagabond, rebel – have seemingly merged into a fused and interwoven character.

I overcame almost all the initial hardships and personally prevailed. The alternation between despair and jubilation, idyllic longing and harsh reality, impetuous gambles and conservative choices that characterized my younger years has been replaced by a sense of calm and belonging, contentment and steadiness, and a hopefully well-managed gambling habit.

As I write the final sentences of this manuscript, I sit at my desk surrounded by three laptop computers in my comfortable apartment high above a lively and multicultural neighborhood – my "home, sweet home" with cheap rent and total silence at night. My day has consisted of student presentations in an online group Zoom session, reading the news about Donald Trump's latest bullshit con artist demagoguery, reviewing and editing some pages, and a walk in the park. I live in a medium-sized city in the European Zone, on the other side of the Atlantic Ocean, away from the America of money, work, media, and consumerist obsessions. I am away from the right-wing faux populism and narcissistic autocracy of "dear leader" Trump.

Looking back on the expectations placed on me by my environment and myself, I feel a sense of triumph regarding what I did with my life. After many travels and varied jobs, I worked for a few decades as a software developer and made enough money to break free. After that, I wrote books on media theory, cultural theory, science fiction theory, and art and technology – publications that garnered academic and public recognition. Last year, I finally got my Ph.D. – a Doctor of Philosophy in "artistic and media research."

I had a first wife for twenty-four years who loved me and a second wife for eight years who loves me now. I have proven myself to be "relationship-capable."

Astonishingly, I became semi-famous as a thinker, a minor intellectual celebrity. After 2012, I changed careers, stopped doing IT consulting work to make money, and started teaching at art and design universities in Germany, Switzerland, and Italy. I have taught nearly one hundred seminars in future design research, transdisciplinary design, post-humanist philosophy, and programming or "Creative Coding" for artists. I have been interviewed on TV and radio shows in several countries: Italy, Germany, Austria, Switzerland, Ireland, the Netherlands, and Estonia. I have been the invited keynote speaker about twenty times at academic, scientific, and business conferences; science fiction fan conventions; and art festivals. I spoke at events solely about my work in front of audiences of five hundred people in Berlin and Milan.

One motivation for my longing for Europe was to attain psychological and existential "individuation" from my parents, to dislodge them from the pre-eminent position they occupied in my psyche as a determining pattern of my behavior and actions.

Why, in my childhood, did I not tell my parents I was unhappy and ask to reverse the skipping twice? One apparent reason is that honest communication in my family was not possible. But another truth is that I was remarkably complicit in my own suffering. I had not thought much about that aspect until now. When my mother told me at age six that I had a uniquely special high IQ, I adopted the idea that I was smarter than everyone else as my identity. I say this now as a deep self-criticism. Yet, on the positive side, I allow myself to say that I did something very different from what others might have done in that situation: I worked actively to transform my self-image of possessing high intelligence into moral intelligence.

A crucial goal was to obtain enough money to enable the practical freedom to be creative. More than anything, as a thinker, I wanted to develop conceptually a coherent political, social, and economic philosophy for articulating how to improve justice and equality in the world beyond capitalism and its known failed alternatives. I developed my intellectual project.

My objective was to be able to support myself (and maybe a wife) by doing something that was not deadly boring and that would not involve killing myself with overwork. Somehow, someday, I would get back to my dream of being a writer. I would write books. I would be an artist – as Camus called himself and others who might follow his credo. I will self-finance to do it. Not with money from my parents. Not with philanthropic "money for artists" from the "culture state." Not with any psychological sense of entitlement.

Beyond all those dreams, I would then, in the future, find a unity between America and Europe. Not the America of America First and Make America Great Again, but the good America I had seen at certain moments of my life, the other America that fulfilled its promise of a just and egalitarian society, the America of the "twenty basic principles of *Star Trek*."[211] Not the Europe of fascism, but the better Europe I had sensed at certain moments, the Europe of the Spanish anarchists, Paris and Prague 1968, and Bologna 1977. Then I would not be guilty of a wasted life. I would have achieved what my mother told me to do when I was six years old.

Then I could have a Happy Death.[212] I could face the hour of my capital punishment with serenity. All that remained to transpire at the moment of the guillotine blade slicing through my neck, severing my head from my body, would be that I accept The End, and that some whom I love, and the world, shall continue without me.

# Notes

1. Paul Auster, *The Music of Chance* (Viking, 1990).
2. Alan N. Shapiro, *Decoding Digital Culture with Science Fiction: Hyper-Modernism, Hyperreality, Posthumanism* (Transcript Verlag, 2024); *Transdisciplinary Design* (Passagen Verlag, 2017); *Die Software der Zukunft* (translated English to German by Marcel Marburger) (Walther König Verlag, 2014); *The Technological Herbarium* (Gianna Maria Gatti, AVINUS Verlag, 2010); *Star Trek: Technologies of Disappearance* (AVINUS Verlag, 2004).
3. George Orwell, *Down and Out in Paris and London* (Victor Gollancz, 1933).
4. Paul Nizan, *Aden Arabie* (foreword by Jean-Paul Sartre) (originally published in French in 1931) (trans. Joan Pinkham) (Monthly Review Press, 1968).
5. Jack Kerouac, *On the Road* (Viking Press, 1957).
6. William Morris, *News From Nowhere, or An Epoch of Rest* (1890).
7. George Orwell, *Keep the Aspidistra Flying* (Victor Gollancz, 1936).
8. Stephen E. Ambrose, *Band of Brothers* (Simon & Schuster, 1992).
9. Jean Baudrillard, *Simulacra and Simulation* (originally published in French in 1981) (trans. Sheila Faria Glaser) (University of Michigan Press, 1994).
10. Meredith L. Clausen, *The Pan Am Building and the Shattering of the Modernist Dream* (The MIT Press, 2006).
11. Jean Baudrillard and Jean Nouvel, *The Singular Objects of Architecture* (originally published in French in 2000) (trans. Robert Bononno) (University of Minnesota Press, 2002); Colin Fournier, "L'architecture de la séduction," in François L'Yvonnet, ed., *Jean Baudrillard: Cahiers de l'Herne* (Éditions de l'Herne, 2004).
12. Franz Kafka, *Give It Up and Other Short Stories* (illustrated by Peter Kuper) (NBM Publishing Company, 1995).
13. Alan N. Shapiro, *A New Alternative to Capitalism: Designing a Moral Economic System with Advanced AI and VR Technologies* (forthcoming, 2026).
14. H.G. Wells, *The War of the Worlds* (William Heinemann, 1898).
15. F. Scott Fitzgerald, *The Great Gatsby* (Charles Scribner's Sons, 1925).
16. Benjamin Fine, *Stretching Their Minds: The Exciting New Approach to the Education of the Gifted Child Pioneered by the Sands Point Country Day School* (E.P. Dutton & Company, 1964).
17. James Fenimore Cooper, *The Last of the Mohicans* (H.C. Carey & I. Lea, 1926).
18. Marjorie Kinnan Rawlings, *The Yearling* (Charles Scribner's Sons, 1938).
19. Mark Twain, *The Adventures of Tom Sawyer* (American Publishing Company, 1876); *Adventures of Huckleberry Finn* (Charles L. Webster and Company, 1884).
20. Marshall McLuhan and Bruce R. Powers, *The Global Village: Transformations in World Life and Media in the 21st Century* (Oxford University Press, 1989).
21. J.D. Salinger, *The Catcher in the Rye* (Little, Brown and Company, 1951).
22. Edgar Allan Poe, *The Narrative of Arthur Gordon Pym of Nantucket* (Harper & Brothers, 1838); Robert Louis Stevenson, *Treasure Island* (Cassell and Company, 1883); Daniel Defoe, *Robinson Crusoe* (William Taylor, 1719); Johan David Wyss, *The Family Robinson Crusoe, or, Journal of a Father Shipwrecked, with his Wife and Children, on an Uninhabited Island* (originally published in German in 1812) (trans. William Godwin and Mary Jane Clairmont) (Juvenile Library, 1816).

23  Paul Goodman, *Growing Up Absurd* (Random House, 1960).
24  Thomas Berger, *Little Big Man* (Dial Press, 1964).
25  Lewis Carroll, *Alice's Adventures in Wonderland* (Macmillan, 1865).
26  Rachel Carson, *Silent Spring* (Houghton Mifflin, 1962).
27  Laurence Wylie, *Village in the Vaucluse* (Harvard University Press, third edition, 2005).
28  John Steinbeck, *The Grapes of Wrath* (Viking Press, 1939); *Of Mice and Men* (Covici Friede, 1937); Arthur Miller, *Death of a Salesman* (Viking Press – play premiered in 1949); *The Crucible* (Viking Press – play premiered in 1953).
29  Richard Wright, *Black Boy* (Harper & Brothers, 1945); Ralph Ellison, *Invisible Man* (Random House, 1952).
30  Alain Touraine, *The May Movement: Revolt and Reform* (originally published in French in 1968) (trans. Leonard F. X. Mayhew) (Random House, 1971); Alfred Willener, *The Action-Image of Society: On Cultural Politicization* (originally published in French in 1970) (trans. A.M. Sheridan Smith) (Routledge Reprint Edition, 2013).
31  Nathan Glazer and Daniel P. Moyniham, *Beyond the Melting Pot: The Negroes, Puerto Ricans, Jews, Italians, and Irish of New York City* (The MIT Press, 1963).
32  Aldous Huxley, *Brave New World* (Chatto & Windus, 1932); Ray Bradbury, *Fahrenheit 451* (Ballantine Books, 1953); *The Martian Chronicles* (Doubleday, 1950); Kurt Vonnegut, *Cat's Cradle* (Holt, Rinehart and Winston, 1963); *Slaughterhouse-Five, or, The Children's Crusade: A Duty-Dance with Death* (Delacorte, 1969).
33  Jack Kerouac, *The Dharma Bums* (Viking, 1958); Richard Brautigan, *Trout Fishing in America* (Four Seasons Foundation, 1967); *A Confederate General from Big Sur* (Grove Press, 1965).
34  Herman Melville, *Billy Budd, Sailor* (University of Chicago Press, 1962); Jack London, *The Call of the Wild* (Macmillan, 1903).
35  Allen Ginsberg, *Howl and Other Poems* (Introduction by William Carlos Williams) (City Lights Books, 1955); Sylvia Plath, *The Colossus and Other Poems* (Heinemann, 1960); R. W. Franklin, ed., *The Poems of Emily Dickinson* (The Belknap Press of Harvard University Press, 1998); Walt Whitman, *Leaves of Grass* (self-published, 1855).
36  William Shakespeare, *The Essential Shakespeare Tragedies Collection: Hamlet, Macbeth, Romeo & Juliet, King Lear, Othello* (2021); Henrik Ibsen, *A Doll's House and Other Plays* (trans. R. Farquharson Sharp) (Penguin, 2016); Arthur Miller, *Timebends: A Life* (Methuen, 1987); George Bernard Shaw, *Man and Superman* (Penguin, 2000).
37  Ralph Waldo Emerson, *Essays (compilation of First and Second Series)* (1841, 1844); Henry David Thoreau, *Walden; or Life in the Woods* (Ticknor and Fields, 1854).
38  Richard Eckaus and Kirit S. Parikh, *Planning for Growth: Multisectoral, Intertemporal Models Applied to India* (The MIT Press, 1968).
39  Paul A. Samuelson, *Economics: An Introductory Analysis* (McGraw-Hill, 1948).
40  John Kenneth Galbraith, *The Affluent Society* (Houghton Mifflin, 1958); *The New Industrial State* (Houghton Mifflin, 1967).
41  Myron Weiner, *The Child and the State in India: Child Labor and Education Policy in Comparative Perspective* (Princeton University Press, 1991).
42  Louis Menand Papers, MC 473, box X. Massachusetts Institute of Technology, Department of Distinctive Collections, Cambridge, Massachusetts.
43  Edward Pincus, *Guide to Filmmaking* (Penguin 1969); Richard Leacock, *The Feeling of Being There – a Filmmaker's Memoir* (Semeïon Editions, 2011).

44  Edward S. Herman and Noam Chomsky, *Manufacturing Consent: The Political Economy of the Mass Media* (New York: Pantheon, 2002).
45  Lorenzo Morris, *Elusive Equality: The Status of Black Americans in Higher Education* (Howard University Press, 1979).
46  W.E.B. Du Bois, *The Souls of Black Folk* (A.C. McClurg & Co., 1903).
47  Thomas S. Kuhn, *The Structure of Scientific Revolutions* (University of Chicago Press, 1962).
48  Jean-Paul Sartre, *Nausea* (originally published in French in 1938) (trans. Robert Baldick) (New Directions, 2013); Albert Camus, *The Stranger* (originally published in French in 1942) (trans. Matthew Ward) (Vintage Book, 1988).
49  Bertrand Russell, *Why I Am Not a Christian and Other Essays on Religion and Related Subjects* (Touchstone Books, 1957).
50  Ludwig Wittgenstein, *Philosophical Investigations* (originally published in German in 1953) (trans. G.E.M. Anscombe) (Pearson, 1973).
51  Bertrand Russell, *Why I Am Not a Christian*.
52  Bertrand Russell, *A History of Western Philosophy and Its Connection with Political and Social Circumstances from the Earliest Times to the Present Day* (Simon and Schuster, 1946).
53  Bertrand Russell, *The Conquest of Happiness* (George Allen & Unwin, 1930); *In Praise of Idleness and Other Essays* (George Allen & Unwin, 1935); *War Crimes in Vietnam* (George Allen & Unwin, 1967).
54  Bertrand Russell, *The Autobiography of Bertrand Russell* (three volumes) (George Allen & Unwin, 1951, 1956, 1969).
55  George Orwell, *Nineteen Eighty-Four* (Secker & Warburg, 1949); *Animal Farm: A Fairy Story* (Secker & Warburg, 1949); *The Road to Wigan Pier* (Victor Gollancz, 1937); *Homage to Catalonia* (Secker & Warburg, 1938); *A Collection of Essays* (Mariner Books Classics, 1970).
56  Fyodor Dostoevsky, *The Gambler and Other Stories* (originally published in Russian in 1863) (trans. Ronald Meyer) (Penguin, 2010); Franz Kafka, *The Trial* (originally published in German in 1925) (trans. Breon Mitchell) (Schocken, 1999); Søren Kierkegaard, *Either/Or: A Fragment of Life* (originally published in Danish in 1843) (trans. Alastair Hannay) (Penguin, 1992); Karl Jaspers, *Philosophy of Existence* (originally published in German in 1938) (trans. Richard F. Grabau) (University of Pennsylvania Press, 1971); Friedrich Nietzsche, *On the Genealogy of Morals* (originally published in German in 1887) (trans. Michael A. Scarpitti) (Penguin, 2013).
57  William Shakespeare, *The Essential Shakespeare Tragedies Collection: Hamlet, Macbeth, Romeo & Juliet, King Lear, Othello*; Herman Melville, *Moby-Dick; or, The Whale* (Harper & Brothers, 1951); Joseph Conrad, *Heart of Darkness* (Blackwood's Magazine, 1899); James Joyce, *Dubliners* (Grant Richards Ltd., 1914); Flannery O'Connor, *A Good Man Is Hard to Find and Other Stories* (Harcourt, Brace and Company, 1955); Virginia Woolf, *Mrs. Dalloway* (Hogarth Press, 1925); Chinua Achebe, *Things Fall Apart* (William Heinemann, 1958).
58  René Descartes, *A Discourse on Method: Meditations and Principles* (originally published in French in 1637) (trans. John Veitch) (The Liberal Arts Press, 1960); Immanuel Kant, *Groundwork of the Metaphysics of Morals* (originally published in German in 1795) (trans. Jens Timmermann) (Cambridge University Press, 2012); David Hume, *A Treatise of Human Nature: Being an Attempt to Introduce the Experimental Method of Reasoning into Moral Subjects* (originally published in 1739-1740) (Penguin, 1986); George Berkeley, *Three Dialogues between Hylas and Philonous* (originally published in 1713) (Hackett Classics, 1979); Michael L. Morgan, ed., *Spinoza Complete Works* (trans. Samuel Shirley) (Hackett

Classics, 2002); Gottfried Wilhelm Leibniz, *Philosophical Essays* (trans. Roger Ariew and Daniel Garber) (Hackett Classics, 1989); Thomas Aquinas, *Selected Writings* (Penguin, 1998); Saint Augustine, *The Confessions* (originally written in 397-400 A.D.) (trans. Henry Chadwick) (Oxford World's Classics, 2009); Willard V.O. Quine, *Word and Object* (The MIT Press, 1960).

59 Dominick LaCapra, *Rethinking Intellectual History: Texts, Contexts, Language* (Cornell University Press, 1983).

60 Franz Kafka, *The Trial*; Albert Camus, *The Stranger*.

61 Albert Camus, *The Myth of Sisyphus and Other Essays* (originally published in French in 1943) (trans. Justin O'Brien) (Vintage Books, 1955).

62 Ibid.; p. 3.

63 Friedrich Nietzsche, *Thus Spoke Zarathustra: A Book for All and None* (originally published in German in 1883, 1892) (trans. R.J. Hollingdale) (Penguin, 1961).

64 Albert Camus, *The Myth of Sisyphus and Other Essays*; p. 5.

65 Ibid; p. 76.

66 Ibid; p. 10.

67 Ibid; p. 16.

68 Albert Camus, *The Rebel: An Essay on Man in Revolt* (originally published in French in 1951) (trans. Anthony Bower) (Vintage, 1992).

69 Ibid; p. 88.

70 Ibid; p. 91.

71 Erich Auerbach, *Mimesis: The Representation of Reality in Western Literature* (originally published in German in 1946) (trans. Willard R. Trask) (Princeton University Press, 1953); Northrup Frye, *Anatomy of Criticism: Four Essays* (Princeton University Press, 1957); György Lukács, *Theory of the Novel: A Historico-Philosophical Essay on the Forms of Great Epic Literature* (originally published in German in 1916-1920) (translated by Anna Bostock) (The MIT Press, 1974); Roland Barthes, *Writing Degree Zero* (originally published in French in 1953) (trans. Annette Lavers) (Vintage, 2010).

72 Charles Darwin, *On the Origin of Species* (John Murray, 1859).

73 Alexis de Tocqueville, *The Old Regime and the French Revolution* (originally published in French in 1856) (trans. Stuart Gilbert) (Anchor, 1983); Joseph de Maistre, *Considerations on France* (originally published in French in 1794) (Cambridge University Press, 1994); Edmund Burke, *Reflections on the Revolution in France* (James Dodsley, 1790).

74 Georg Wilhelm Friedrich Hegel, *The Phenomenology of Spirit* (originally published in German in 1807) (trans. Terry Pinkard) (Cambridge University Press, 2018); Karl Marx, *The German Ideology, Including Theses on Feuerbach* (originally written in German in 1845-1846) (Prometheus Books, 1988); Stendhal, *The Red and the Black* (originally published in French in 1930) (trans. Roger Gard) (Penguin, 2002); Fyodor Dostoevsky, *The Gambler and Other Stories*; Gustave Flaubert, *Madame Bovary* (originally published in French in 1857) (trans. Lydia Davis) (Penguin, 2011); Auguste Comte, *A General View of Positivism* (originally published in French in 1948) (trans. J.J. Bridges) (Cambridge University Press, 2009); John Stuart Mill, *On Liberty* (originally published in 1859) (CreateSpace, 2014); Emile Durkheim, *Suicide: A Study in Sociology* (originally published in French in 1897) (trans. John A. Spaulding and George Simpson) (The Free Press, 1951); Friedrich Nietzsche, *On the Genealogy of Morals*.

75 Karl Marx, *Capital: A Critique of Political Economy, Vol. 1* (trans. Ben Fowkes) (Penguin, 1990).

76   Karl Marx, *Economic and Philosophic Manuscripts of 1844* (trans. Martin Milligan) (Lawrence & Wishart, 1977).
77   Theodor W. Adorno, *Minima Moralia: Reflections from Damaged Life* (originally published in German in 1951) (trans. Dennis Redmond) (Verso, 2020); Herbert Marcuse, *One-Dimensional Man: Studies in the Ideology of Advanced Industrial Society* (Beacon Press, 1964); Walter Benjamin, *Illuminations: Essays and Reflections* (edited by Hannah Arendt) (Schocken, 1969).
78   Gustave Flaubert, *Sentimental Education: The Story of a Young Man* (originally published in French in 1869) (trans. Raymond N. MacKenzie) (University of Minnesota Press, 2024); *Flaubert's Dictionary of Accepted Ideas* (originally published in French in 1911-1913) (trans. Jacques Barzun) (New Directions, 1968).
79   Gustave Flaubert, *Bouvard and Pécuchet* (originally published in French in 1881) (trans. Mark Polizzotti) (Dalkey Archive, 2005).
80   Sigmund Freud, *The Interpretation of Dreams* (originally published in German in 1899) (trans. Abraham Arden Brill) (The Macmillan Company, 1913); C. Wright Mills, ed., *From Max Weber* (trans. Hans Gerth) (Free Press, 1946).
81   Ludwig Wittgenstein, *Philosophical Investigations*; Thomas Mann, *Death in Venice and Other Stories* (originally published in German in 1912) (trans. Kenneth Burke) (Alfred A. Kopf, 1925); Virginia Woolf, *To the Lighthouse* (Hogarth Press, 1927); Jean-Paul Sartre, *Nausea*; Albert Camus, *The Plague* (originally published in French in 1947) (trans. Stuart Gilbert) (Penguin, 1960); Herbert Marcuse, *Eros and Civilization: A Philosophical Inquiry into Freud* (Beacon Press, 1955); Michel Foucault, *Madness and Civilization: A History of Insanity in the Age of Reason* (originally published in French in 1961) (trans. Richard Howard) (Vintage, 1988); Jürgen Habermas, *Toward a Rational Society: Student Protest, Science, and Politics* (trans. Jeremy J. Shapiro) (Beacon Press, 1971).
82   Albert Einstein, *Relativity: The Special and the General Theory* (Fingerprint Publishing, 2017); Niels Bohr, *Atomic Theory and the Description of Nature: Four Essays with an Introductory Survey* (Cambridge University Press, 2011).
83   István Mészáros, *Marx's Theory of Alienation* (The Merlin Press, 2006); Walter A. Kaufmann, *Nietzsche: Philosopher, Psychologist, Antichrist* (Princeton University Press, 2013); Mikhail Bakhtin, *Problems of Dostoevsky's Poetics* (originally published in Russian in 1929) (trans. R. William Rotsel) (University of Michigan Press, 1973); Jean-Paul Sartre, *The Words* (originally published in French in 1963) (trans. Bernard Frechtman) (Vintage, 1964); Donald Lazere, *The Unique Creation of Albert Camus* (Yale University Press, 1973).
84   Ayn Rand, *The Fountainhead* (Bobbs Merrill, 1943).
85   Rupert Roopnaraine, *The Sky's Wild Noise: Selected Essays* (Peepal Tree Press, 2013).
86   Charles Dickens, *The Pickwick Papers* (Chapman & Hall, 1837).
87   Jean-Paul Sartre, *The Family Idiot: Gustave Flaubert, 1821–1857, An Abridged Edition* (originally published in French in 1971-1972) (trans. Carol Cosman) (University of Chicago Press, 2023).
88   Mikhail Bakhtin, *Problems of Dostoevsky's Poetics*; *Rabelais and His World* (trans. Tvorchestvo Fransua Rable) (The MIT Press, 1968); V.N. Voloshinov, *Marxism and the Philosophy of Language* (originally written in Russian in the late 1920s) (trans. L. Matejka and I.R. Titunik) (Harvard University Press, 1973); György Lukács, *History and Class Consciousness: Studies in Marxist Dialectics* (originally published in German in 1923) (trans. Rodney Livingstone) (The MIT Press, 1972); Lucien Goldmann, *The Hidden God: a study of tragic*

*vision in the Pensées of Pascal and the tragedies of Racine* (originally published in French in 1955) (trans. Philip Thody) (Routledge, 1964).
89  Louis-Ferdinand Céline, *Journey to the End of the Night* (originally published in French in 1932) (trans. John H. P. Marks) (Chatto and Windus, 1934); André Gide, *The Counterfeiters* (originally published in French in 1925) (trans. Dorothy Bussy) (Alfred A. Knopf, 1927); Italo Svevo, *The Confessions of Zeno* (originally published in Italian in 1923) (trans. Beryl De Zoete) (Vintage, 1958).
90  Walter Rodney, *How Europe Underdeveloped Africa* (Introduction by Vincent Harding) (Bogle-l'Ouverture Publications, 1972).
91  John Weiss, *The Making of Technological Man: The Social Origins of French Engineering Education* (The MIT Press, 1982).
92  Hannah Arendt, *Eichmann in Jerusalem: A Report on the Banality of Evil* (Viking, 1963); Erich Maria Remarque, *All Quiet on the Western Front* (originally published in German in 1929) (trans. A.W. Ween) (Ballantine, 1987); Carlo Levi, *Christ Stopped at Eboli: The Story of a Year* (originally published in Italian in 1945) (trans. Frances Frenaye) (Picador, 2020).
93  Eldon Kenworthy, *America/Américas: Myth in the Making of U.S. Policy Toward Latin America* (Penn State University Press, 1995).
94  Ricardo Rojo, *My Friend Che* (Dial Press, 1968).
95  Murray Bookchin, *Post-Scarcity Anarchism* (originally published in 1971) (Edinburgh: AK Press, 2004); E.F. Schumacher, *Small is Beautiful: A Study of Economics As If People Mattered* (originally published in 1973) (Harper Perennial, 2010).
96  Jean-Paul Sartre, *The Communists and Peace, with a Reply to Claude Lefort* (originally published in French in July 1952 in *Les Temps Modernes*) (trans. Martha H. Fletcher) (G. Braziller, 1968).
97  Antonio Gramsci, *Prison Notebooks* (originally published in Italian in 1929) (trans. Joseph Anthony Buttigieg II) (Columbia University Press, 1991-2011).
98  Jean Baudrillard, *The Mirror of Production* (originally published in French in 1973) (translated by Mark Poster) (Telos Press, 1975).
99  Jean Baudrillard, *The Consumer Society: Myths and Structures* (originally published in French in 1970) (trans. Chris Turner) (Sage, 1998).
100  Guy Debord, *The Society of the Spectacle* (originally published in French in 1967) (trans. Donald Nicholson-Smith) (Zone Books, 1994); Guy Debord, *The Society of the Spectacle* (trans. Ken Knabb) (Rebel Press, n.d.).
101  Ibid.; theses 2, 9.
102  Ibid.; theses 1, 2, 5, 6.
103  Rem Koolhaas, *Constant: New Babylon* (Hatje Cantz, 2016).
104  Martin Jay, *The Dialectical Imagination: A History of the Frankfurt School and the Institute of Social Research, 1923-1950* (University of California Press, 1973).
105  Francis M. Du Mont, *French Grammar* (Barnes & Noble Outline series, 1969).
106  Petite Collection Maspero (Éditions Maspero) – Book Series.
107  Marc Silver, *Arguing the Case: Language and Play in Argumentation* (Adams Press, 1996).
108  Maurizio Calvesi, *Avanguardia di massa: Compaiono gli indiani metropolitani* (Postmedia Books, 2018).
109  Gianfranco Sanguinetti, *Truthful Report on the Last Chances to Save Capitalism in Italy* (trans. Bill Brown) (Colossal Books, 2013).

110 Guy Debord, *Comments on the Society of the Spectacle* (trans. Malcolm Imrie) (Verso, 1998).
111 Lea Melandi, *L'infamia originaria: Facciamola finita col cuore e la politica* (Manifestolibri, 2020); Cornelius Castoriadis, *The Imaginary Institution of Society* (originally published in French in 1975) (trans. Kathleen Blamey) (Polity, 1987).
112 Lea Melandi, *L'infamia originaria: Facciamola finita col cuore e la politica*.
113 Albert Camus, "Create Dangerously," in *Resistance, Rebellion, and Death* (trans. Justin O'Brien) (Vintage, 1995).
114 Umberto Eco, *Semiotics and the Philosophy of Language* (Indiana University Press, 1986).
115 Umberto Eco, *Travels in Hyperreality* (trans. William Weaver) (Harcourt Brace Jovanovich, 1995).
116 Franco "Bifo" Berardi, *Finalmente il cielo è caduto sulla terra* (Squilibri Edizioni, 1978).
117 Thomas Mann, *Death in Venice* (originally published in German in 1912) (trans. Stanley Applebaum) (Dover, 1995).
118 Birgit Haustedt, *Rilke's Venice* (originally published in German in 2006) (trans. Stephen Brown) (Haus Publishing, 2008); pp. 47,51; Rainer Maria Rilke, *Stories of God* (originally published in German in 1900) (trans. M.D. Herder) (Norton, 1932); p. 68.
119 *Rilke's Venice*; p. 52; *Stories of God*; p. 71.
120 Rainer Maria Rilke, *Gesammelte Werke (Gedichte)* (Penguin Random House Verlagsgruppe, 2000).
121 Albert Camus, "Nuptials at Tipasa," in *Lyrical and Critical Essays* (trans. Ellen Conroy Kennedy) (Vintage Books, 1970); pp. 65-72.
122 Ibid; p. 66.
123 Ibid; p. 71.
124 Albert Camus, "Return to Tipasa," in *The Myth of Sisyphus and Other Essays*; pp. 195-204.
125 Ibid; p. 197.
126 Ibid; p. 202.
127 Ibid; p. 203.
127 Karl Marx and Friedrich Engels, *The Communist Manifesto* (originally published in German in 1848) (trans. Samuel Moore) (Progress Publishers, 1977).
128 Stanley Aronowitz, *False Promises: The Shaping of American Working Class Consciousness* (Duke University Press, 1991).
129 Stuart Ewen, *Captains of Consciousness: Advertising and the Social Roots of the Consumer Culture* (Basic Books, 2001).
130 Barbara Ehrenreich, *Nickel and Dimed: On (Not) Getting By in America* (Picador Paper, 2021).
131 Antonio Gramsci, "The Formation of the Intellectuals," in *Selections from the Prison Notebooks* (trans. Quentin Hoare and Geoffrey Nowell Smith) (International Publishers, 1971).
132 Paul Avrich, *Kronstadt, 1921* (Princeton University Press, 2014).
133 Peter Hudis, *Marx's Concept of the Alternative to Capitalism* (AAKAR, 2016).
134 George Orwell, *Down and Out in Paris and London* (Victor Gollancz, 1933); Chapter One.
135 Ibid; Chapters Ten and Eleven.
136 Lawrence Ferlinghetti, *A Coney Island of the Mind* (New Directions, 1958); Allen Ginsberg, *Howl and Other Poems* (City Lights, 1956).

137 *A Survey of American Gambling Attitudes and Behavior* (University of Michigan Institute for Social Research); p. 87.
138 Françoise Castel, Robert Castel, and Anne Lovell, *The Psychiatric Society* (Columbia University Press, 1982).
139 Jean Baudrillard, "The Beaubourg Effect: Implosion and Deterrence," in *Simulacra and Simulation* (originally published in French in 1977); pp. 61-73.
140 Egeria Di Nallo, *Indiani in Città* (Nuova Universale Cappelli, 1977).
141 Claude Lefort, *The Political Forms of Modern Society: Bureaucracy, Democracy, Totalitarianism* (The MIT Press, 1986).
142 Nancy Mangabeira Unger, *Fundamentos Filosóficos do Pensamento Ecológico* (Loyola, 1992).
143 Cornelius Castoriadis, Claude Lefort, et. al. *Socialisme ou Barbarie: An Anthology* (ERIS, 2019).
144 Italo Svevo, *Una vita* (Libreria Editrice Ettore Vram, 1892); *Senilità* (Libreria Editrice Ettore Vram, 1927); *La Coscienza di Zeno* (Cappelli, 1930).
145 Hunter S. Thompson, *Fear and Loathing in Las Vegas: A Savage Journey to the Heart of the American Dream* (Random House, 1972).
146 Eric Bentley, *A Time To Live & A Time To Die* (Broadway Play Publishing, 1967); Bertolt Brecht, *Mother Courage and Her Children* (originally written in German in 1938-1939) (trans. Eric Bentley) (Grove Press, 1991).
147 *The Collected Poetry of W.H. Auden* (Random House, 1945).
148 *Telos*; quarterly journal, 1968 to present.
149 Paul Piccone, *Italian Marxism* (University of California Press, 1983).
150 Ernst Bloch, *The Principle of Hope* (originally published in German in 1954) (trans. Neville Plaice, Stephen Plaice, and Paul Knight) (The MIT Press, 1986); Henri Lefebvre, *Everyday Life in the Modern World* (originally published in French in 1968) (trans. Sacha Rabinovitch) (Penguin, 1971).
151 Andrew Arato, *From Neo-Marxism to Democratic Theory: Essays on the Critical Theory of Soviet-type Societies* (Routledge, 2016).
152 Gershom Scholem, *On the Kabbalah and Its Symbolism* (originally published in German in 1960) (translated by Ralph Manheim) (Schocken Books, 1965); pp. 211-12.
153 Dennis Wrong, *Power: Its Forms, Bases, and Uses* (Transaction Publishers, 1980).
154 Eliot Freidson, *Profession of Medicine: A Study of the Sociology of Applied Knowledge* (New York University Press, 1970).
155 Juan Corradi, *South of the Crisis: A Latin American Perspective on the Late Capitalist World* (Anthem Press, 2010).
156 Wolf Heydebrand and Carroll Serron, *Rationalizing Justice: The Political Economy of Federal District Courts* (State University of New York Press, 1990).
157 Claude Lefort, *Le Temps present. Écrits 1945-2005* (Belin, 2007).
158 Claude Lefort, *Le Travail de l'œuvre, Machiavel* (Gallimard, 1972); *Un Homme en trop. Essai sur l'archipel du goulag de Soljénitsyne* (Le Seuil, 1975); *Sur une colonne absente. Autour de Merleau-Ponty* (Gallimard, 1978); *L'Invention démocratique. Les limites de la domination totalitaire* (Fayard, 1981).
159 Karel Kosik, *Dialectics of the Concrete: A Study on Problems of Man and World* (D. Reidel, 1976).
160 Edgar Morin, Claude Lefort, and Cornelius Castoriadis, *Mai 68, La Brèche* (Fayard, 2008).

161 Victor Brombert, *The Novels of Flaubert: A Study of Themes and Techniques* (Princeton University Press, 1966).
162 Alan N. Shapiro, "Confidence in Public Institutions," unpublished paper, 1982.
163 Christopher Williams, *The Song Remains the Same*, unpublished Ph.D. dissertation manuscript, 1995.
164 Erving Goffman, *The Presentation of Self in Everyday Life* (Doubleday, 1959).
165 Alain Robbe-Grillet, *For a New Novel* (originally published in French in 1963) (trans. Richard Howard) (Northwestern University Press, 1992).; *Last Year at Marienbad* (originally written in French in 1961) (trans. Richard Howard) (Grove, 1962).
166 Nathalie Sarraute, *The Planetarium* (originally published in French in 1959) (trans. Maria Jolas) (Dalkey Archive Essentials, 2022); Michel Butor, *Passing Time* (originally published in French in 1956) (trans. Jean Stewart) (Simon & Schuster, 1960); Claude Simon, *The Grass* (originally published in French in 1958) (trans. Richard Howard) (George Braziller, 1960).
167 Alain Robbe-Grillet, *Projet pour une révolution à New York* (Les Éditions de Minuit, 1970).
168 Karl Marx, *Economic and Philosophic Manuscripts of 1844* (trans. Martin Milligan) (Lawrence & Wishart, 1977); *The German Ideology, Including Theses on Feuerbach*.
169 Walter Benjamin, *Illuminations: Essays and Reflections*; Theodor W. Adorno, *Minima Moralia: Reflections from Damaged Life*; Max Horkheimer, *Eclipse of Reason* (Oxford University Press, 1947); Herbert Marcuse, *One-Dimensional Man: Studies in the Ideology of Advanced Industrial Society*; Jürgen Habermas, *Toward a Rational Society: Student Protest, Science, and Politics*.
170 Emile Durkheim, *Suicide: A Study in Sociology*.
171 Karl Mannheim, *Ideology and Utopia: An Introduction to the Sociology of Knowledge* (originally published in German in 1929) (trans. Louis Wirth and Edward Shils) (Martino Fine Books, 2015).
172 Friedrich Nietzsche, *On the Genealogy of Morals*; Michel Foucault, *Madness and Civilization: A History of Insanity in the Age of Reason*; Jacques Derrida, *Writing and Difference* (originally published in French in 1967) (trans. Alan Bass) (University of Chicago Press, 2017).
173 Alvin Gouldner, *The Future of Intellectuals and the Rise of the New Class* (Continuum, 1979); Gyorgy Konrad and Ivan Szelenyi, *The Intellectuals on the Road to Class Power* (trans. Andrew Arato and Richard E. Allen) (Harcourt Brace Jovanovich, 1979).
174 Christopher Lasch, "The Corruption of Sports," in *The New York Review of Books*; April 28, 1977.
175 Georg Simmel, *The Philosophy of Money* (originally published in German in 1900) (trans. David Frisby and Tom Bottomore) (Routledge, 2004).
176 Alan N. Shapiro, "American *Samizdat*: From Cultural Citizenship to Cultural Critique," unpublished manuscript, 1985.
177 Jean Baudrillard, *Simulacra and Simulation*; Roland Barthes, *Mythologies* (originally published in French in 1957) (trans. Annette Lavers) (Farrar, Straus & Giroux, 1972); Jacques Derrida, *Writing and Difference*; David Riesman with Nathan Glazer and Reuel Denney, *The Lonely Crowd: A Study of the Changing American Character* (Yale University Press, 1950); Daniel Boorstin, *The Image: A Guide to Pseudo-Events in America* (Harper Colophon, 1962); Christopher Lasch, *The Culture of Narcissism: American Life in an Age of Diminishing Expectations* (W.W. Norton, 1979).

# NOTES

178 Herbert Marcuse, *One-Dimensional Man: Studies in the Ideology of Advanced Industrial Society.*
179 Irving L. Janus, *Groupthink* (Houghton Mifflin, 1982); George Orwell, *Nineteen Eighty-Four.*
180 Richard Sennett, *The Fall of Public Man* (Knopf, 1977).
181 Richard Sennett, *The Corrosion of Character, The Personal Consequences Of Work In the New Capitalism* (Norton, 1998).
182 *Telos*; quarterly journal, 1968 to present; Jean Baudrillard, *The Mirror of Production*; *For a Critique of the Political Economy of the Sign* (originally published in French in 1972) (trans. Charles Levin) (St. Lous: Telos Press, 1981).
183 Paul Piccone, unpublished letter to Alan N. Shapiro, May 24, 1985.
184 Alan N. Shapiro, "American *Samizdat*: From Cultural Citizenship to Cultural Critique."
185 Gene Schoor (Alan N. Shapiro ghostwriter), *The Complete Dodgers Record Book* (Facts on File, 1984); *The Complete Yankees Record Book* (Facts on File, 1984); *The Complete Cardinals Record Book* (Facts on File, 1984).
186 Daniel Cohn-Bendit and Gabriel Cohn-Bendit, *Obsolete Communism: The Left-Wing Alternative* (originally published in French in 1968) (trans. Arnold Pomerans) (McGraw Hill, 1969).
187 Jean Baudrillard, *The Ecstasy of Communication* (originally published in French in 1987) (trans. Bernard Schutze and Caroline Schutze) (Semiotext(e), 1988).
188 Robert Sedgewick, *Algorithms* (Addison-Wesley, 1978).
189 Brian W. Kernighan and Dennis M. Ritchie, *The C Programming Language* (Prentice-Hall, 1978).
190 Achim Szepanski, *Die Ekstase der Spekulation: Kapitalismus im Zeitalter der Katastrophe* (Galerie der abseitigen Künste, 2023).
191 Guy Debord, *Comments on the Society of the Spectacle.*
192 Jean Baudrillard, *America* (originally published in French in 1986) (trans. Chris Turner) (Verso, 1988); pp. 127-28.
193 Gregory Bateson, *Steps to an Ecology of Mind: Collected Essays in Anthropology, Psychiatry, Evolution, and Epistemology* (University Of Chicago Press, 1972).
194 William Shakespeare, *The Merchant of Venice* (Cambridge University Press, 2014).
195 Robert Venturi, Denise Scott Brown, and Steven Izenour, *Learning from Las Vegas: The Forgotten Symbolism of Architectural Form* (Cambridge MA: The MIT Press, 1972).
196 Jean Baudrillard, *Fatal Strategies* (originally published in French in 1983) (trans. Philip Beitchman and W.G.J. Niesluchowski) (Semiotext(e), 1983).
197 Jean Baudrillard, *Simulacra and Simulation*; pp. 12-13.
198 Jean-Paul Sartre, *Existentialism is a Humanism* (originally published in French in 1945) (trans. Carol Macomber) (Yale University Press, 2007).
199 Roland Barthes, *Mythologies.*
200 Jean Baudrillard, *Fatal Strategies.*
201 Chris Kraus, "*Chance*-Event," in Peter Gente, Barbara Könches and Peter Weibel, eds., *Philosophie und Kunst Jean Baudrillard: Eine Hommage zu seinem 75. Geburtstag* (translated from English to German) (Berlin, Merve Verlag, 2005); pp. 295-310.
202 Howard Rollin Patch, *The Goddess Fortuna in Medieval Literature* (Harvard University Press, 1927).

203 *The Defeat of a Renaissance Intellectual: Selected Writings of Francesco Guicciardini* (trans. Carlo Celli) (Pennsylvania State University Press, 2019).
204 Niccolò Machiavelli, *The Prince* (originally published in Italian in 1532) (trans. Tim Parks) (Penguin, 2009).
205 Walter Benjamin, *The Arcades Project* (originally published in German in 1982) (trans. Howard Eiland and Kevin McLaughlin) (Belknap Press, 2002).
206 Fyodor Dostoevsky, *The Gambler and Other Stories*.
207 Paul Auster, *The Music of Chance*.
208 Paul Auster, *The Red Notebook* (Faber and Faber, 1995).
209 Jean Baudrillard, *Seduction* (originally published in French in 1979) (trans. Brian Singer) (New World Perspectives, 1990); p. 133.
210 For the analysis bringing together Auster's *The Music of Chance* and Baudrillard's *Seduction*, I am indebted to Eyal Dotan, "The Game of Late Capitalism: Gambling and Ideology in *The Music of Chance*," *Mosaic: An Interdisciplinary Critical Journal* (March 2000).
211 Alan N. Shapiro, "*Star Trek* Twenty Basic Principles," in *Star Trek: Technologies of Disappearance* (Berlin: AVINUS Verlag, 2004) pp. 25-34.
212 Albert Camus, *A Happy Death* (originally published in French in 1971) (trans. Richard Howard) (Vintage 1973).

# Studies in Criticality

*Series Editor*
*Shirley R. Steinberg*

Counterpoints publishes the most compelling and imaginative books being written in Education and Cultural Studies today. Grounded on the theoretical advances in critical theory, feminism, and postcolonialism in the last two decades of the twentieth century, Counterpoints engages the meaning of these innovations in various forms of educational expression. Committed to the proposition that theoretical literature should be accessible to a variety of audiences, the series insists that its authors avoid esoteric and jargonistic languages that transform educational scholarship into an elite discourse for the initiated. Scholarly work matters only to the degree it affects consciousness and practice at multiple sites. The editorial policy of *Counterpoints* is based on these principles and the ability of scholars to break new ground, to open new conversations, to go where educators have never gone before.

For additional information about this series or for the submission of manuscripts, please contact:

>Shirley R. Steinberg, Series Editor
>msgramsci@gmail.com

To order other books in this series, please contact our Customer Service Department:

>peterlang@presswarehouse.com (within the U.S.)
>orders@peterlang.com (outside the U.S.)

Or browse online by series:

>www.peterlang.com